# Applied Engineering Mathematics

# Applied Engineering Mathematics

**Xin-She Yang**

University of Cambridge, Cambridge, United Kingdom

Viva Books

New Delhi | Mumbai | Chennai | Kolkata | Bangalore | Hyderabad | Kochi

**First Indian Edition 2008**

**VIVA BOOKS PRIVATE LIMITED**

- 4737/23 Ansari Road, Daryaganj, New Delhi 110 002
  E-mail: vivadelhi@vivagroupindia.net, Tel. 42242200
- Plot No. 76, Service Industries, Shirvane, Sector 1, Nerul, Navi Mumbai 400 706
  E-mail: vivamumbai@vivagroupindia.net, Tel. 27721273, 27721274
- Jamals Fazal Chambers, 26 Greams Road, Chennai 600 006
  E-mail: vivachennai@vivagroupindia.net, Tel. 28290304, 28294241
- B-103, Jindal Towers, 21/1A/3 Darga Road, Kolkata 700 017
  E-mail: vivakolkata@vivagroupindia.net, Tel. 22836381, 22816713
- 7, Sovereign Park Apartments, 56-58, K. R. Road, Basavanagudi, Bangalore 560 004
  E-mail: vivabangalore@vivagroupindia.net, Tel. 26607409, 26607410
- 101-102 Mughal Marc Apartments, 3-4-637 to 641, Narayanguda, Hyderabad 500 029
  E-mail: vivahyderabad@vivagroupindia.net, Tel. 27564481, 27564482
- First Floor, Beevi Towers, SRM Road, Kaloor, Kochi 682 018, Kerala
  E-mail: vivakochi@vivagroupindia.net, Tel: 0484-2403055, 2403056

*www.vivagroupindia.com*

*Published by arrangement with*

Cambridge International Science Publishing
7, Meadow Walk, Great Abington
Cambridge CB1 6AZ, UK



ISBN-10: 81-309-0953-7
ISBN-13: 978-81-309-0953-0

Published by Vinod Vasishtha for Viva Books Private Limited, 4737/23, Ansari Road, Daryaganj, New Delhi-110 002. Printed and bound by Anand Sons, Delhi - 110 092.

# Preface

Engineering mathematics including numerical methods and application is the essential part of key problem-solving skills for engineers and scientists. Modern engineering design and process modelling require both mathematical analysis and computer simulations. Vast literature exists on engineering mathematics, mathematical modelling and numerical methods. The topics in engineering mathematics are very diverse and the syllabus of mathematics itself is evolving. Therefore, there is a decision to select the topics and limit the number of chapters so that the book remains concise and yet comprehensive enough to include all the important mathematical methods and popular numerical methods.

This book endeavors to strike a balance between mathematical and numerical coverage of a wide range of mathematical methods and numerical techniques. It strives to provide an introduction, especially for undergraduates and graduates, to engineering mathematics and its applications. Topics include advanced calculus, ordinary differential equations, partial differential equations, vector and tensor analysis, calculus of variations, integral equations, the finite difference method, the finite volume method, the finite element method, reaction-diffusion system, and probability and statistics. The book also emphasizes the application of important mathematical methods with dozens of worked examples. The applied topics include elasticity, harmonic motion, chaos, kinematics, pattern formation and hypothesis testing. The book can serve as a textbook in engineering mathematics, mathematical modelling, and scientific computing.

Xin-She Yang
Cambridge, 2007

# Acknowledgements

First and foremost, I would like to thank my mentors, tutors and colleagues: Prof. A C Fowler and Prof. C J Mcdiarmid at Oxford University, Dr J M Lees and Dr C T Morley at Cambridge University, Prof. A C McIntosh, Prof. J Brindley, Prof. R W Lewis, Prof. D T Gethin, and Prof. Andre Revil for their help and encouragement. I also thank Dr G. Parks, Dr T. Love, Dr S. Guest, Dr K. Seffen, and many colleagues for their inspiration. I thank many of my students, especially Hugo Whittle and Charles Pearson, at Cambridge University who have indirectly tried some parts of this book and gave their valuable suggestions.

I also would like to thank my publisher, Dr Victor Riecansky, for his kind help and professional editing.

Last but not least, I thank my wife, Helen, and son, Young, for their help and support.

Xin-She Yang

## About the Author

Xin-She Yang received his D.Phil in applied mathematics from the University of Oxford. He is currently a research fellow at the University of Cambridge. Dr Yang has published extensively in international journals, book chapters, and conference proceedings. His research interests include asymptotic analysis, bioinspired algorithms, combustion, computational engineering, engineering optimization, solar eclipses, scientific programming and pattern formation. He is also the author of a book entitled: "An Introduction to Computational Engineering with Matlab", published in 2006 by Cambridge International Science Publishing Ltd.

# Contents

# Chapter 1

# Calculus

The preliminary requirements for this book are the pre-calculus foundation mathematics. We assume that the readers are familiar with these preliminaries, and readers can refer to any book that is dedicated to these topics. Therefore, we will only review some of the basic concepts of differentiation and integration.

## 1.1 Differentiations

### 1.1.1 Definition

For a known function or a curve $y = f(x)$ as shown in Figure 1.1, the slope or the gradient of the curve at the point $P(x, y)$ is defined as

$$\frac{dy}{dx} \equiv \frac{df(x)}{dx} \equiv f'(x) = \lim_{\Delta x \to 0} \frac{f(x + \Delta x) - f(x)}{\Delta x}, \tag{1.1}$$

on the condition that there exists such a limit at $P$.

This gradient or limit is the first derivative of the function $f(x)$ at $P$. If the limit does not exist at a point $P$, then we say that the function is non-differentiable at $P$. By convention, the limit of the infinitesimal change $\Delta x$ is denoted as the differential $dx$. Thus, the above definition can also be written as

$$dy = df = \frac{df(x)}{dx} dx = f'(x)dx, \tag{1.2}$$

which can be used to calculate the change in $dy$ caused by the small change of $dx$. The primed notation $'$ and standard notation $\frac{d}{dx}$ can be used interchangeably, and the choice is purely out of convenience.

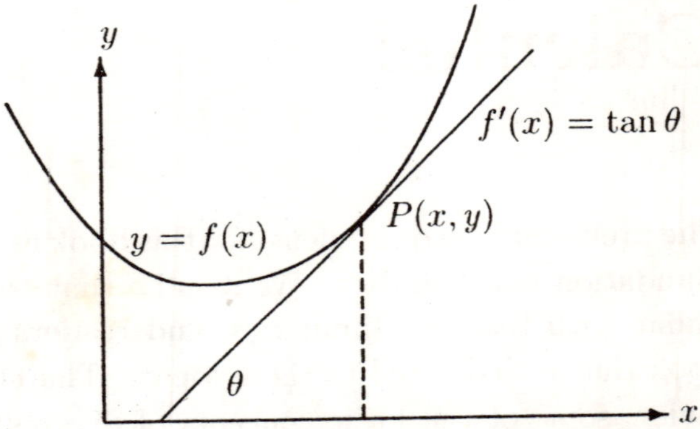


Figure 1.1: Gradient of a curve

The second derivative of $f(x)$ is defined as the gradient of $f'(x)$, or

$$\frac{d^2y}{dx^2} \equiv f''(x) = \frac{df(x)}{dx}. \tag{1.3}$$

The higher derivatives can be defined in a similar manner. Thus,

$$\frac{d^3y}{dx^3} \equiv f'''(x) = \frac{df''(x)}{dx}, \quad ..., \quad \frac{d^ny}{dx^n} \equiv f^{(n)} = \frac{df^{(n-1)}}{dx}. \tag{1.4}$$

### 1.1.2 Differentiation Rules

If a more complicated function $f(x)$ can be written as a product of two simpler functions $u(x)$ and $v(x)$, we can derive a differentiation rule using the definition from the first principles. Using

$$\frac{f(x+\Delta x)-f(x)}{\Delta x}=\frac{u(x+\Delta x)v(x+\Delta x)-u(x)v(x)}{\Delta x},$$

and subtracting and adding $-u(x+\Delta x)v(x)+u(x+\Delta x)v(x)$ [$=0$] terms, we have

$$\frac{df}{dx}=\frac{d[u(x)v(x)]}{dx}$$

$$=\lim_{\Delta x\to 0}[u(x+\Delta x)\frac{v(x+\Delta x)-v(x)}{\Delta x}+v(x)\frac{u(x+\Delta x)-u(x)}{\Delta x}]$$

$$=u(x)\frac{dv}{dx}+\frac{du}{dx}v(x), \tag{1.5}$$

which can be written in a contract form using primed notations

$$f'(x)=(uv)'=u'v+uv'. \tag{1.6}$$

If we differentiate this equation again and again, we can generalize this rule, we finally get the Leibnitz's Theorem for differentiations

$$\frac{d^n(uv)}{dx^n}=u^{(n)}v+nu^{(n-1)}v'+...+\left(\begin{array}{c}n\\r\end{array}\right)u^{(n-r)}v^{(r)}$$

$$+...+uv^{(n)}, \tag{1.7}$$

where the coefficients are the same as the binomial coefficients

$${}^nC_r\equiv\left(\begin{array}{c}n\\r\end{array}\right)=\frac{n!}{r!(n-r)!}. \tag{1.8}$$

If a function $f(x)$ [for example, $f(x)=e^{x^n}$] can be written as a function of another function $g(x)$, or $f(x)=f[g(x)]$ [for example, $f(x)=e^{g(x)}$ and $g(x)=x^n$], then we have

$$f'(x)=\lim_{\Delta x\to 0}\frac{\Delta f}{\Delta g}\frac{\Delta g}{\Delta x}, \tag{1.9}$$

which leads to the following chain rule

$$f'(x) = \frac{df}{dg}\frac{dg}{dx}, \tag{1.10}$$

or

$$\{f[g(x)]\}' = f'[g(x)] \cdot g'(x). \tag{1.11}$$

In our example, we have $f'(x) = (e^{x^n})' = e^{x^n} n x^{n-1}$.

If one use $1/v$ instead of $v$ in the equation (1.6) and $(1/v)' = -v'/v^2$, we have the following differentiation rule for quotients:

$$(\frac{u}{v})' = \frac{u'v - uv'}{v^2}. \tag{1.12}$$

---

□ ***Example 1.1:*** *The derivative of $f(x) = \sin(x)e^{-\cos(x)}$ can be obtained using the combination of the above differentiation rules.*

$$f'(x) = [\sin(x)]' e^{-\cos(x)} + \sin(x)[e^{-\cos(x)}]'$$
$$= \cos(x)e^{-\cos(x)} + \sin(x)e^{-\cos(x)}[-\cos(x)]'$$
$$= \cos(x)e^{-\cos(x)} + \sin^2(x)e^{-\cos(x)}.$$

□

---

The derivatives of various functions are listed in Table 1.1.

### 1.1.3 Implicit Differentiation

The above differentiation rules still apply in the case when there is no simple explicit function form $y = f(x)$ as a function of $x$ only. For example, $y + \sin(x)\exp(y) = 0$. In this case, we can differentiate the equation term by term with respect to $x$ so that we can obtain the derivative $dy/dx$ which is in general a function of both $x$ and $y$.

---

□ ***Example 1.2:*** *Find the derivative $\frac{dy}{dx}$ if $y^2 + \sin(x)e^y = \cos(x)$. Differentiating term by term with respect to $x$, we have*

$$2y\frac{dy}{dx} + \cos(x)e^y + \sin(x)e^y\frac{dy}{dx} = -\sin(x),$$

$$\frac{dy}{dx} = -\frac{\cos(x)e^y + \sin(x)}{2y + \sin(x)e^y}.$$

□

---

Table 1.1: First Derivatives

| $f(x)$ | $f'(x)$ |
|---|---|
| $x^n$ | $nx^{n-1}$ |
| $e^x$ | $e^x$ |
| $a^x (a > 0)$ | $a^x \ln a$ |
| $\ln x$ | $\frac{1}{x}$ |
| $\log_a x$ | $\frac{1}{x \ln a}$ |
| $\sin x$ | $\cos x$ |
| $\cos x$ | $-\sin x$ |
| $\tan x$ | $\sec^2 x$ |
| $\sin^{-1} x$ | $\frac{1}{\sqrt{1-x^2}}$ |
| $\cos^{-1} x$ | $-\frac{1}{\sqrt{1-x^2}}$ |
| $\tan^{-1} x$ | $\frac{1}{1+x^2}$ |
| $\sinh x$ | $\cosh x$ |
| $\cosh x$ | $\sinh x$ |

## 1.2 Integrations

### 1.2.1 Definition

Integration can be viewed as the inverse of differentiation. The integration $F(x)$ of a function $f(x)$ satisfies

$$\frac{dF(x)}{dx} = f(x), \tag{1.13}$$

or

$$F(x) = \int_{x_0}^{x} f(\xi) d\xi, \tag{1.14}$$

where $f(x)$ is called the integrand, and the integration starts from $x_0$ (arbitrary) to $x$. In order to avoid any potential confusion, it is conventional to use a dummy variable (say, $\xi$) in the integrand. As we know, the geometrical meaning of the first derivative is the gradient of the function $f(x)$ at a point $P$, the geometrical representation of an integral $\int_a^b f(\xi)d\xi$ (with lower integration limit $a$ and upper integration limit $b$) is the area under the curve $f(x)$ enclosed by $x$-axis in the region $x \in [a, b]$. In this case, the integral is called a definite integral as the limits are given. For the definite integral, we have

$$\int_a^b f(x)dx = \int_{x_0}^b f(x)dx - \int_{x_0}^a f(x)dx = F(b) - F(a). \quad (1.15)$$

The difference $F(b) - F(a)$ is often written in a compact form $F|_a^b \equiv F(b) - F(a)$. As $F'(x) = f(x)$, we can also write the above equation as

$$\int_a^b f(x)dx = \int_a^b F'(x)dx = F(b) - F(a). \quad (1.16)$$

Since the lower limit $x_0$ is arbitrary, the change or shift of the lower limit will lead to an arbitrary constant $c$. When the lower limit is not explicitly given, the integral is called an indefinite integral

$$\int f(x)dx = F(x) + c, \quad (1.17)$$

where $c$ is the constant of integration.

The integrals of some of the common functions are listed in Table 1.2.

### 1.2.2 Integration by Parts

From the differentiation rule $(uv)' = uv' + u'v$, we have

$$uv' = (uv)' - u'v. \quad (1.18)$$

Integrating both sides, we have

$$\int u\frac{dv}{dx}dx = uv - \int \frac{du}{dx}vdx, \quad (1.19)$$

in the indefinite form. It can also be written in the definite form as

$$\int_a^b u\frac{dv}{dx}dx = [uv]\Big|_a^b + \int_a^b v\frac{du}{dx}dx. \quad (1.20)$$

Table 1.2: Integrals

| $f(x)$ | $\int f(x)$ |
|---|---|
| $x^n (n \neq -1)$ | $\frac{x^{n+1}}{n+1}$ |
| $\frac{1}{x}$ | $\ln \lvert x \rvert$ |
| $e^x$ | $e^x$ |
| $\sin x$ | $-\cos x$ |
| $\cos x$ | $\sin x$ |
| $\frac{1}{a^2+x^2}$ | $\frac{1}{a}\tan^{-1}\frac{x}{a}$ |
| $\frac{1}{a^2-x^2}$ | $\frac{1}{2a}\ln\frac{a+x}{a-x}$ |
| $\frac{1}{x^2-a^2}$ | $\frac{1}{2a}\ln\frac{x-a}{x+a}$ |
| $\frac{1}{\sqrt{a^2-x^2}}$ | $\sin^{-1}\frac{x}{a}$ |
| $\frac{1}{\sqrt{x^2+a^2}}$ | $\ln(x+\sqrt{x^2+a^2})$ [or $\sinh^{-1}\frac{x}{a}$] |
| $\frac{1}{\sqrt{x^2-a^2}}$ | $\ln(x+\sqrt{x^2-a^2})$ [or $\cosh^{-1}\frac{x}{a}$] |
| $\sinh x$ | $\cosh x$ |
| $\cosh x$ | $\sinh x$ |
| $\tanh x$ | $\ln \cosh x$ |

The integration by parts is a very powerful method for evaluating integrals. Many complicated integrands can be rewritten as a product of two simpler functions so that their integrals can easily obtained using integration by parts.

---

□ ***Example 1.3:*** *The integral of $I = \int x \ln x \, dx$ can be obtained by setting $v' = x$ and $u = \ln x$. Hence, $v = \frac{x^2}{2}$ and $u' = \frac{1}{x}$. We now have*

$$I = \int x \ln x dx = \frac{x^2 \ln x}{2} - \int \frac{x^2}{2}\frac{1}{x} dx$$

$$= \frac{x^2 \ln x}{2} - \frac{x^2}{4}.$$

□

---

Other important methods of integration include the substitution and reduction methods. Readers can refer any book that is dedicated to advanced calculus.

### 1.2.3 Taylor Series and Power Series

From

$$\int_a^b f(x)dx = F(b) - F(a), \tag{1.21}$$

and $\frac{dF}{dx} = F' = f(x)$, we have

$$\int_{x_0}^{x_0+h} f'(x)dx = f(x_0 + h) - f(x_0), \tag{1.22}$$

which means that

$$f(x_0 + h) = f(x_0) + \int_{x_0}^{x_0+h} f'(x)dx. \tag{1.23}$$

If $h$ is not too large or $f'(x)$ does not vary dramatically, we can approximate the integral as

$$\int_{x_0}^{x_0} f'(x)dx \approx f'(x_0)h. \tag{1.24}$$

Thus, we have the first-order approximation to $f(x_0 + h)$

$$f(x_0 + h) \approx f(x_0) + hf'(x_0). \tag{1.25}$$

This is equivalent to say, any change from $x_0$ to $x_0+h$ is approximated by a linear term $hf'(x_0)$. If we repeat the procedure for $f'(x)$, we have

$$f'(x_0 + h) \approx f'(x_0) + hf''(x_0), \tag{1.26}$$

which is a better approximation than $f'(x_0 + h) \approx f'(x_0)$. Following the same procedure for higher order derivatives, we can reach the $n$-th order approximation

$$f(x_0 + h) = f(x_0) + hf'(x_0) + \frac{h^2}{2!}f''(x_0) + \frac{h^3}{3!}f'''(x_0)$$

$$+ ... + \frac{h^n}{n!} f^{(n)}(x_0) + R_{n+1}(h), \tag{1.27}$$

where $R_{n+1}(h)$ is the error of this approximation and the notation means that the error is about the same order as $n+1$-th term in the series. This is the well-known Taylor theorem and it has many applications. In deriving this formula, we have implicitly assumed that all the derivatives $f'(x), f''(x), ..., f^{(n)}(x)$ exist. In almost all the applications we meet, this is indeed the case. For example, $\sin(x)$ and $e^x$, all the orders of the derivatives exist. If we continue the process to infinity, we then reach the infinite power series and the error $\lim_{n\to\infty} R_{n+1} \to 0$ if the series converges. The end results are the Maclaurin series. For example,

$$e^x = 1 + x + \frac{x^2}{2!} + ... + \frac{x^x}{x!} .... \quad (x \in \mathcal{R}), \tag{1.28}$$

$$\sin x = x - \frac{x^3}{3!} + \frac{x^5}{5!} - .... \quad (x \in \mathcal{R}), \tag{1.29}$$

$$\cos x = 1 - \frac{x^2}{2!} + \frac{x^4}{4!} - .... \quad (x \in \mathcal{R}), \tag{1.30}$$

and

$$\ln(1+x) = x - \frac{x^2}{2} + \frac{x^3}{3} - \frac{x^4}{4} + \frac{x^5}{5} - .... \quad (-1 < x \le 1). \tag{1.31}$$

## 1.3 Partial Differentiation

### 1.3.1 Partial Differentiation

The differentiation defined above is for function $f(x)$ which has only one independent variable $x$, and the gradient will generally depend on the location $x$. For functions $f(x, y)$ of two variables $x$ and $y$, their gradient will depend on both $x$ and $y$ in general. In addition, the gradient or rate of change will also depend on the direction (along $x$-axis or $y$-axis or any other directions). For example, the function $f(x, y) = xy$ shown in Figure 1.2 has different gradients at $(0, 0)$ along $x$-axis and $y$-axis. The

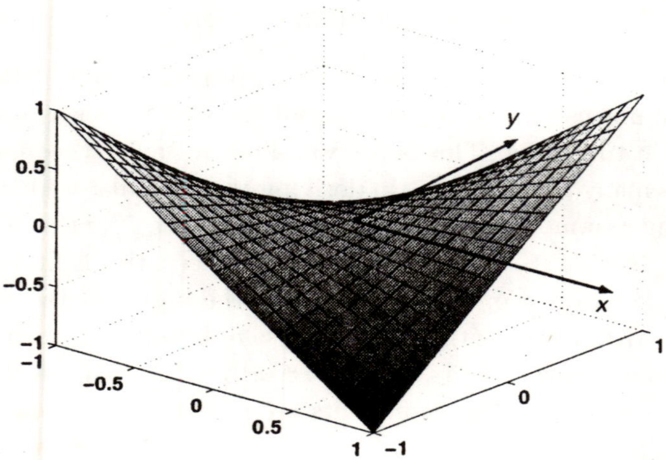


Figure 1.2: Variation of $f(x, y) = xy$.

gradients along the positive $x$- and $y$- directions are called the partial derivatives respect to $x$ and $y$, respectively. They are denoted as $\frac{\partial f}{\partial x}$ and $\frac{\partial f}{\partial y}$, respectively.

The partial derivative of $f(x, y)$ with respect to $x$ can be calculated assuming that $y$ =constant. Thus, we have

$$\frac{\partial f(x,y)}{\partial x} \equiv f_x \equiv \frac{\partial f}{\partial x}|_y$$

$$= \lim_{\Delta x \to 0, y=const} \frac{f(x+\Delta x, y) - f(x,y)}{\Delta x}. \tag{1.32}$$

Similarly, we have

$$\frac{\partial f(x,y)}{\partial y} \equiv f_y \equiv \frac{\partial f}{\partial y}|_x$$

$$= \lim_{\Delta y \to 0, x=const} \frac{f(x, y+\Delta y) - f(x,y)}{\Delta y}. \tag{1.33}$$

The notation $\frac{\partial}{\partial x}|_y$ emphasizes the fact that $y$ is held constant. The subscript notation $f_x$ (or $f_y$) emphasizes the derivative is carried out with respect to $x$ (or $y$). Mathematicians like to use the subscript forms as they are simpler notations and can be easily generalized. For example,

$$f_{xx} = \frac{\partial^2 f}{\partial x^2}, \qquad f_{xy} = \frac{\partial^2 f}{\partial x \partial y}. \tag{1.34}$$

Since $\Delta x \Delta y = \Delta y \Delta x$, we have $f_{xy} = f_{yx}$.

---

□ ***Example 1.4:*** *The first partial derivatives of* $f(x, y) = xy + \sin(x)e^{-y}$ *are*

$$f_x = \frac{\partial f}{\partial x} = y + \cos(x)e^{-y}, \qquad f_y = \frac{\partial f}{\partial y} = x - \sin(x)e^{-y}.$$

*The second partial derivative of* $f(x, y)$ *is*

$$f_{xx} = -\sin(x)e^{-y}, \qquad f_{yy} = \sin(x)e^{-y},$$

*and*

$$f_{xy} = f_{yx} = 1 - \cos(x)e^{-y}.$$

□

---

For any small change $\Delta f = f(x+\Delta x, y+\Delta y) - f(x, y)$ due to $\Delta x$ and $\Delta y$, the total infinitesimal change $df$ can be written as

$$df = \frac{\partial f}{\partial x}dx + \frac{\partial f}{\partial y}dy. \tag{1.35}$$

If $x$ and $y$ are functions of another independent variable $\xi$, then the above equation leads to the following chain rule

$$\frac{df}{d\xi} = \frac{\partial f}{\partial x}\frac{dx}{d\xi} + \frac{\partial f}{\partial y}\frac{dy}{d\xi}, \tag{1.36}$$

which is very useful in calculating the derivatives in parametric form or for change of variables. If a complicated function $f(x)$ can be written in terms of simpler functions $u$ and $v$ so that $f(x) = g(x, u, v)$ where $u(x)$ and $v(x)$ are known functions of $x$, then we have the generalized chain rule

$$\frac{dg}{dx} = \frac{\partial g}{\partial x} + \frac{\partial g}{\partial u}\frac{du}{dx} + \frac{\partial g}{\partial v}\frac{dv}{dx}. \tag{1.37}$$

The extension to functions of more than two variables is straightforward. For a function $p(x, y, z, t)$ such as the pressure in a fluid, we have the total differential as

$$df = \frac{\partial p}{\partial t}dt + \frac{\partial p}{\partial x}dx + \frac{\partial p}{\partial y}dy + \frac{\partial p}{\partial z}dz. \tag{1.38}$$

### 1.3.2 Differentiation of an Integral

When differentiating an integral

$$\Phi(x) = \int_a^b \phi(x, y)dy, \tag{1.39}$$

with fixed integration limits $a$ and $b$, we have

$$\frac{\partial \Phi(x)}{\partial x} = \int_a^b \frac{\partial \phi(x, y)}{\partial x}dy. \tag{1.40}$$

When differentiating the integrals with the limits being functions of $x$,

$$I(x) = \int_{v(x)}^{u(x)} \psi(x, \tau)d\tau = \Psi[x, u(x)] - \Psi[x, v(x)], \tag{1.41}$$

the following formula is useful:

$$\frac{dI}{dx} = \int_{v(x)}^{u(x)} \frac{\partial \psi}{\partial x}d\tau + [\psi(x, u(x))\frac{du}{dx} - \psi(x, v(x))\frac{dv}{dx}]. \tag{1.42}$$

This formula can be derived using the chain rule

$$\frac{dI}{dx} = \frac{\partial I}{\partial x} + \frac{\partial I}{\partial u}\frac{du}{dx} + \frac{\partial I}{\partial v}\frac{dv}{dx}, \tag{1.43}$$

where $\frac{\partial I}{\partial u} = \psi(x, u(x))$ and $\frac{\partial I}{\partial v} = -\psi(x, v(x))$.

## 1.4 Multiple Integrals

### 1.4.1 Multiple Integrals

As the integration of a function $f(x)$ corresponds to the area enclosed under the function between integration limits, this can extend to the double integral and multiple integrals. For a function $f(x, y)$, the double integral is defined as

$$F = \int_{\Omega} f(x, y)dA, \tag{1.44}$$

where $dA$ is the infinitesimal element of the area, and $\Omega$ is the region for integration. The simplest form of $dA$ is $dA = dxdy$ in Cartesian coordinates. In order to emphasize the double integral in this case, the integral is often written as

$$I = \iint_{\Omega} f(x, y)dxdy. \tag{1.45}$$

---

□ ***Example 1.5:*** *The area moment of inertia of a thin rectangular plate, with the length $2a$ and the width $2b$, is defined by*

$$I = \iint_{\Omega} y^2 dS = \iint_{\Omega} y^2 dxdy.$$

*The plate can be divided into four equal parts, and we have*

$$I = 4\int_0^a [\int_0^b y^2 dy]dx = 4\int_0^a \frac{1}{3}b^3 dx$$

$$= \frac{4b^3}{3}\int_0^a dx = \frac{4ab^3}{3}.$$

□

---

### 1.4.2 Jacobian

Sometimes it is necessary to change variables when evaluating an integral. For a simple one-dimensional integral, the change of variables from $x$ to a new variable $v$ (say) leads to $x = x(v)$. This is relatively simple as $dx = \frac{dx}{dv}dv$, and we have

$$\int_{x_a}^{x_b} f(x)dx = \int_a^b f(x(v))\frac{dv}{dx}dv. \tag{1.46}$$

where the integration limits change so that $x(a) = x_a$ and $x(b) = x_b$. Here the extra factor $dx/dv$ in the integrand is referred to as the Jacobian.

For a double integral, it is more complicated. Assuming $x = x(\xi, \eta), y = y(\xi, \eta)$, we have

$$\iint f(x, y)dxdy = \iint f(\xi\eta)|J|d\xi d\eta, \tag{1.47}$$

where $J$ is the Jacobian. That is

$$J \equiv \frac{\partial(x, y)}{\partial(\xi, \eta)}$$

$$= \begin{vmatrix} \frac{\partial x}{\partial \xi} & \frac{\partial x}{\partial \eta} \\ \frac{\partial y}{\partial \xi} & \frac{\partial y}{\partial \eta} \end{vmatrix} = \begin{vmatrix} \frac{\partial x}{\partial \xi} & \frac{\partial y}{\partial \xi} \\ \frac{\partial x}{\partial \eta} & \frac{\partial y}{\partial \eta} \end{vmatrix}. \tag{1.48}$$

The notation $\partial(x, y)/\partial(\xi, \eta)$ is just a useful shorthand. This equivalent to say that the change of the infinitesimal area $dA = dxdy$ becomes

$$dxdy = |\frac{\partial(x, y)}{\partial(\xi, \eta)}|d\xi d\eta = |\frac{\partial x}{\partial \xi}\frac{\partial y}{\partial \eta} - \frac{\partial x}{\partial \eta}\frac{\partial y}{\partial \xi}|d\xi d\eta. \tag{1.49}$$

---

□ ***Example 1.6:*** *When transforming from $(x, y)$ to polar coordinates $(r, \theta)$, we have the following relationships*

$$x = r\cos\theta, \qquad y = r\sin\theta.$$

*Thus, the Jacobian is*

$$J = \frac{\partial(x, y)}{\partial(r, \theta)} = \frac{\partial x}{\partial r}\frac{\partial y}{\partial \theta} - \frac{\partial x}{\partial \theta}\frac{\partial y}{\partial r}$$

$$= \cos\theta \times r\cos\theta - (-r\sin\theta) \times \sin\theta = r[\cos^2\theta + \sin^2\theta] = r.$$

*Thus, an integral in $(x, y)$ will be transformed into*

$$\iint \phi(x, y)dxdy = \iint \phi(r\cos\theta, r\sin\theta)rdrd\theta.$$

□

---

In a similar fashion, the change of variables in triple integrals gives

$$V = \iiint_\Omega \phi(x, y, z)dxdydz = \iint_\omega \psi(\xi, \eta, \zeta)|J|d\xi d\eta d\zeta, \tag{1.50}$$

and

$$J \equiv \frac{\partial(x,y,z)}{\partial(\xi,\eta,\zeta)} = \begin{vmatrix} \frac{\partial x}{\partial \xi} & \frac{\partial y}{\partial \xi} & \frac{\partial z}{\partial \xi} \\ \frac{\partial x}{\partial \eta} & \frac{\partial y}{\partial \eta} & \frac{\partial z}{\partial \eta} \\ \frac{\partial x}{\partial \zeta} & \frac{\partial y}{\partial \zeta} & \frac{\partial z}{\partial \zeta} \end{vmatrix}. \tag{1.51}$$

For cylindrical polar coordinates $(r, \phi, z)$ as shown in Figure 1.3, we have

$$x = r\cos\phi, \qquad y = r\sin\phi, \qquad z = z. \tag{1.52}$$

The Jacobian is therefore

$$J = \frac{\partial(x,y,z)}{\partial(r,\phi,z)} = \begin{vmatrix} \cos\phi & \sin\phi & 0 \\ -r\sin\phi & r\cos\phi & 0 \\ 0 & 0 & 1 \end{vmatrix} = r. \tag{1.53}$$

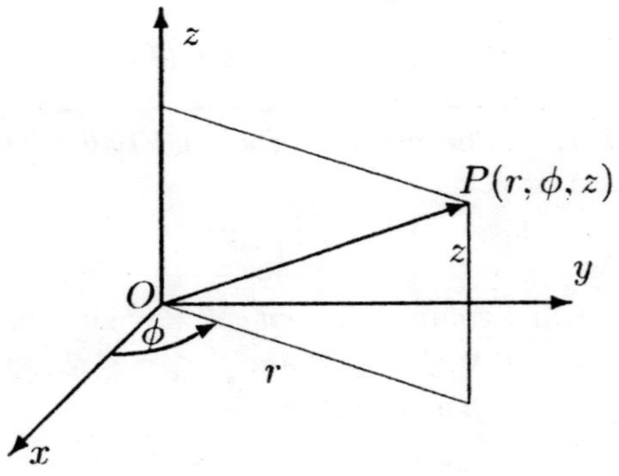


Figure 1.3: Cylindrical polar coordinates.

For spherical polar coordinates $(r, \theta, \phi)$ as shown in Figure 1.4, where $\theta$ is the zenithal angle between the $z$-axis and the position vector $\mathbf{r}$, and $\phi$ is the azimuthal angle, we have

$$x = r\sin\theta\cos\phi, \qquad y = r\sin\theta\sin\phi, \qquad z = r\cos\theta. \tag{1.54}$$

Therefore, the Jacobian is

$$J = \begin{vmatrix} \sin\theta\cos\phi & \sin\theta\sin\phi & \cos\theta \\ r\cos\theta\cos\phi & r\cos\theta\sin\phi & -r\sin\theta \\ -r\sin\theta\sin\phi & r\sin\theta\cos\phi & 0 \end{vmatrix} = r^2\sin\theta. \tag{1.55}$$

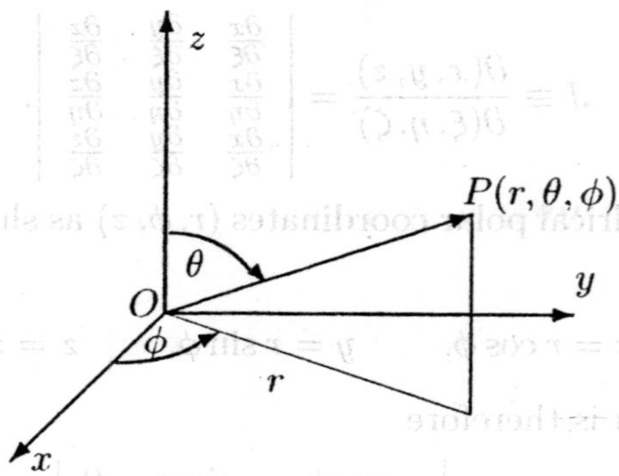


Figure 1.4: Spherical polar coordinates.

Thus, the volume element change in the spherical system is

$$dxdydz = r^2 \sin\theta dr d\theta d\phi. \tag{1.56}$$

---

□ ***Example 1.7:*** *The volume of a solid ball with a radius $R$ is defined as*

$$V = \iiint_\Omega dV.$$

*Since the infinitesimal volume element $dV = r^2 \sin(\theta) dr d\theta d\phi$ in spherical coordinates $r \geq 0, 0 \leq \theta \leq \pi$ and $0 \leq \phi \leq 2\pi$, the ball can be divided into two equal parts so that*

$$V = 2\int_0^R \{\int_0^{\pi/2} \sin\theta [\int_0^{2\pi} d\phi] d\theta\} dr$$

$$= 2\int_0^R \{2\pi \int_0^{\pi/2} \sin(\theta) d\theta\} dr$$

$$= 4\pi \int_0^R r^2 dr = \frac{4\pi}{3} R^3.$$

□

---

## 1.5 Some Special Integrals

Some integrals appear so frequently in engineering mathematics that they deserve special attention. Most of these special integrals are also called special functions as they have certain varying parameters or integral limits. We only discuss four of the most common integrals here.

### 1.5.1 Asymptotic Series

Before we discuss any special functions, let us digress first to introduce the asymptotic series and order notations because they will be used to study the behaviours of special functions. Loosely speaking, for two functions $f(x)$ and $g(x)$, if

$$\frac{f(x)}{g(x)} \to K, \qquad x \to x_0, \tag{1.57}$$

where $K$ is a finite, non-zero limit, we write

$$f = O(g). \tag{1.58}$$

The big $O$ notation means that $f$ is asymptotically equivalent to the order of $g(x)$. If the limit is unity or $K = 1$, we say $f(x)$ is order of $g(x)$. In this special case, we write

$$f \sim g, \tag{1.59}$$

which is equivalent to $f/g \to 1$ and $g/f \to 1$ as $x \to x_0$. Obviously, $x_0$ can be any value, including 0 and $\infty$. The notation $\sim$ does not necessarily mean $\approx$ in general, though they might give the same results, especially in the case when $x \to 0$ [for example, $\sin x \sim x$ and $\sin x \approx x$ if $x \to 0$].

When we say $f$ is order of 100 (or $f \sim 100$), this does not mean $f \approx 100$, but it can mean that $f$ is between about 50 to 150. The small $o$ notation is used if the limit tends to 0. That is

$$\frac{f}{g} \to 0, \qquad x \to x_0, \tag{1.60}$$

or

$$f = o(g). \tag{1.61}$$

If $g > 0$, $f = o(g)$ is equivalent to $f \ll g$. For example, for $\forall x \in \mathcal{R}$, we have $e^x \approx 1 + x + O(x^2) \approx 1 + x + \frac{x^2}{2} + o(x)$.

Another classical example is the Stirling's asymptotic series for factorials

$$n! \sim \sqrt{2\pi n}(\frac{n}{e})^n, \qquad n \gg 1. \tag{1.62}$$

In fact, it can be expanded into more terms

$$n! \sim \sqrt{2\pi n}(\frac{n}{e})^n(1 + \frac{1}{12n} + \frac{1}{288n^2} - \frac{139}{51480n^3} - ...). \tag{1.63}$$

This is a good example of asymptotic series. For standard power expansions, the error $R_k(h^k) \to 0$, but for an asymptotic series, the error of the truncated series $R_k$ decreases and gets smaller compared with the leading term [here $\sqrt{2\pi n}(n/e)^n$]. However, $R_n$ does not necessarily tend to zero. In fact, $R_2 = \frac{1}{12n} \cdot \sqrt{2\pi n}(n/e)^n$ is still very large as $R_2 \to \infty$ if $n \gg 1$. For example, for $n = 100$, we have $n! = 9.3326 \times 10^{157}$, while the leading approximation is $\sqrt{2\pi n}(n/e)^n = 9.3248 \times 10^{157}$. The difference between these two values is $7.7740 \times 10^{154}$, which is still very large, though three orders smaller than the leading approximation.

### 1.5.2 Gaussian Integrals

The Gaussian integral appears in many situations in engineering mathematics and statistics. It can be defined by

$$I(\alpha) = \int_{-\infty}^{\infty} e^{-\alpha x^2} dx. \tag{1.64}$$

In order to evaluate the integral, let us first evaluate $I^2$. Since the Gaussian integral is a definite integral and must give a constant value, we can change the dummy variable as we wish. We have

$$I^2 = [\int_{-\infty}^{\infty} e^{-\alpha x^2} dx]^2 = \int_{-\infty}^{\infty} e^{-\alpha x^2} dx \int_{-\infty}^{\infty} e^{-\alpha y^2} dy$$

$$= \int_{-\infty}^{\infty} \int_{-\infty}^{\infty} e^{-\alpha(x^2+y^2)} dxdy. \quad (1.65)$$

Changing into the polar coordinates $(r, \theta)$ and noticing $r^2 = x^2 + y^2$ and $dxdy = rdrd\theta$, we have

$$I^2 = \int_0^{\infty} dr \int_0^{2\pi} re^{-\alpha r^2} d\theta$$

$$= 2\pi \int_0^{\infty} \frac{1}{\alpha} e^{-\alpha r^2} d(\alpha r^2) = \frac{\pi}{\alpha}. \quad (1.66)$$

Therefore,

$$I(\alpha) = \int_{-\infty}^{\infty} e^{-\alpha x^2} dx = \sqrt{\frac{\pi}{\alpha}}. \quad (1.67)$$

Since $\alpha$ is a parameter, we can differentiate both sides of this equation with respect to $\alpha$, and we have

$$\int_{-\infty}^{\infty} x^2 e^{-\alpha x^2} dx = \frac{1}{2\alpha} \sqrt{\frac{\pi}{\alpha}}. \quad (1.68)$$

By differentiating both sides of the Gaussian integral (equation 1.67) $n$ times with respect to $\alpha$, and we get the generalized Gaussian integral $I_n$

$$I_n = \int_{-\infty}^{\infty} x^{2n} e^{-\alpha x^2}$$

$$= \frac{(-1)^n \cdot 1 \cdot 3 \cdot 5 \cdots (2n-1)}{2^n} \sqrt{\frac{\pi}{\alpha^{2n+1}}}, \quad (1.69)$$

where $n > 0$ is an integer.

For a special case when $\alpha = \frac{1}{\sqrt{2\sigma^2}}$ and $n = 0$, the equation (1.67) can be rearranged as

$$\int_{\infty}^{\infty} f(x, \sigma) dx = 1, \qquad f(x) = \frac{1}{\sqrt{2\pi\sigma^2}} e^{-\frac{x^2}{2\sigma^2}}. \quad (1.70)$$

The function $f(x, \sigma)$ is a zero-mean Gaussian probability function. As $\sigma \to 0$, $f(x) \to \delta(x)$ where $\delta(x)$ is the Dirac $\delta$-function which is defined by

$$\delta(x) \neq 0 \text{ (at } x = 0), \text{ but } \delta(x) = 0, \text{ for } x \neq 0, \quad (1.71)$$

and

$$\int_{\infty}^{\infty} \delta(x)dx = 1. \tag{1.72}$$

It has an interesting property that

$$\int f(x)\delta(x-\beta)dx = f(\beta), \tag{1.73}$$

where $f(x)$ is a function.

### 1.5.3 Error Functions

The error function, which appears frequently in heat conduction and diffusion problems, is defined by

$$\text{erf}(x) = \frac{2}{\sqrt{\pi}} \int_0^x e^{-\eta^2} d\eta. \tag{1.74}$$

Its complementary error function is defined by

$$\text{erfc}(x) = 1 - \text{erf}(x) = \frac{2}{\sqrt{\pi}} \int_x^{\infty} e^{-t^2} dt. \tag{1.75}$$

The error function is an odd function: $\text{erf}(-x) = -\text{erf}(x)$. Using the results from the Gaussian integral

$$\int_{-\infty}^{\infty} e^{-\eta^2} d\eta = \sqrt{\pi}, \tag{1.76}$$

together with the basic definition, we have $\text{erf}(0) = 0$, and $\text{erf}(\infty) = 1$. Both the error function and its complementary function are shown in Figure 1.5.

The error function cannot be easily evaluated in closed form. Using Taylor series for the integrand

$$e^{-\eta^2} = 1 - \eta^2 + \frac{1}{2}\eta^4 - \frac{1}{6}\eta^6 + ..., \tag{1.77}$$

and integrating term by term, we have

$$\text{erf}(x) = \frac{2}{\sqrt{\pi}}[x - \frac{x^3}{3} + \frac{x^5}{10} - \frac{x^7}{42} + ...], \tag{1.78}$$

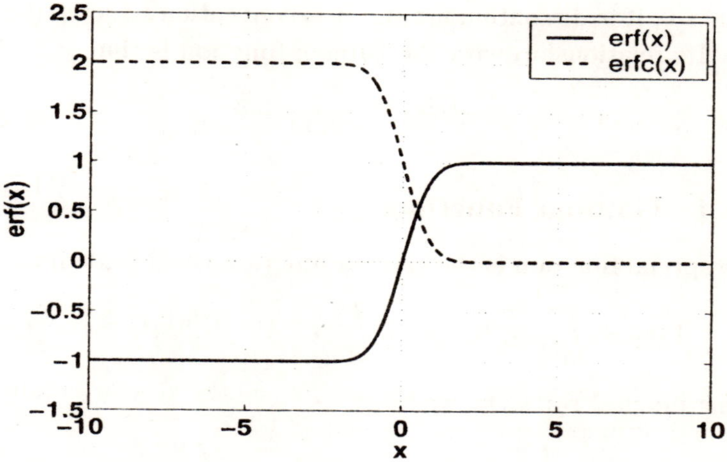


Figure 1.5: Error functions.

or

$$\operatorname{erf}(x) = \frac{2}{\sqrt{\pi}} \sum_{n=0}^{\infty} \frac{(-1)^n}{2n+1} \frac{x^{2n+1}}{n!}. \tag{1.79}$$

The integrals of the complementary function are defined by

$$\operatorname{ierfc}(x) = \int_x^{\infty} \operatorname{erfc}(\eta) d\eta, \tag{1.80}$$

and

$$\mathrm{i^n erfc}(x) = \int_x^{\infty} \mathrm{i^{n-1} erfc}(\eta) d\eta. \tag{1.81}$$

Using integration by parts, we can prove the following asymptotic series

$$\operatorname{erf}(x) \sim 1 - \frac{e^{-x^2}}{x\sqrt{\pi}}, \quad (x \to \infty). \tag{1.82}$$

On the other hand, if we replace $x$ in the error function by $\beta x$, we have

$$\lim_{\beta \to \infty} \frac{1}{2}[1 + \operatorname{erf}(\beta x)] \to H(x). \tag{1.83}$$

where $H(x)$ is a Heaviside function or a unit step function which is defined by

$$H(x) = 1 \ \ (\text{for } \ x > 0), \qquad H(x) = 0 \ \ (\text{for } \ x < 0). \tag{1.84}$$

At $x = 0$, it is discontinuous and it is convention to set $H(0) = 1/2$. Its relationship with the Dirac $\delta$-function is that

$$\frac{dH(x)}{dx} = \delta(x). \tag{1.85}$$

### 1.5.4 Gamma Functions

The special function is the gamma function which is defined by

$$\Gamma(x) = \int_0^\infty t^{x-1}e^{-t}dt = \int_0^\infty e^{-t+(x-1)\ln t}dt. \tag{1.86}$$

Using integral by parts, we have

$$\Gamma(x+1) = \int_0^\infty t^x e^{-t}dt = -t^x e^{-t}\Big|_0^\infty + \int_0^\infty x t^{x-1}e^{-t}dt$$
$$= x\Gamma(x). \tag{1.87}$$

When $x = 1$, we have

$$\Gamma(1) = \int_0^\infty e^{-t}dt = -e^{-t}\Big|_0^\infty = 1. \tag{1.88}$$

The variation of $\Gamma(x)$ is shown in Figure 1.6.

If $x = n$ is an integer ($n \in \mathcal{N}$), then $\Gamma(n+1) = n!$. That is to say,

$$n! = \Gamma(n+1) = \int_0^\infty e^{n\ln t - t}dt. \tag{1.89}$$

The integrand $f(n,t) = \exp[n\ln t - t]$ reaches a maximum value at

$$\frac{\partial f}{\partial t} = 0, \qquad \text{or} \qquad t = n. \tag{1.90}$$

The maximum is $f_{max} = \exp[n\ln n - n]$. Thus, we now can set $t = n + \tau = n(1+\zeta)$ so that $\tau = n\zeta$ varies around $n$ and $\zeta$ around 0. For $n \gg 1$, we have

$$n! = \int_{-\infty}^\infty e^{\{n\ln[n(1+\zeta)] - n(1+\zeta)\}}d\tau$$
$$= \int_{-\infty}^\infty e^{\{(n\ln n - n) + n[\ln(1+\zeta) - \zeta]\}}d\tau, \tag{1.91}$$

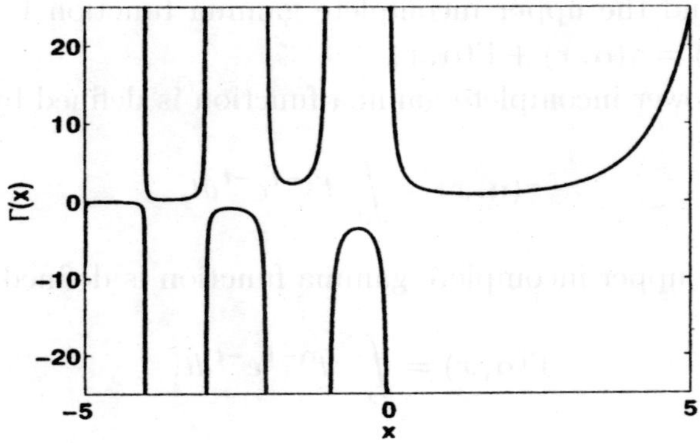


Figure 1.6: Variation of $\Gamma(x)$.

where we have used $\ln[n(1+\zeta)] = \ln n + \ln(1+\zeta)$. The integration limits for $\tau = n\zeta$ (not $\zeta$) are from $-\infty$ to $\infty$. Using

$$\ln(1+\zeta) = \zeta - \frac{\zeta^2}{2} + \frac{\zeta^3}{3} - ..., \tag{1.92}$$

we have

$$n! = e^{n \ln -n} \int_{-\infty}^{\infty} e^{-\frac{\tau^2}{2n}} d\tau. \tag{1.93}$$

From the Gaussian integral with $\alpha = 1/(2n)$

$$\int_{-\infty}^{\infty} e^{-\alpha\tau^2} d\tau = \sqrt{\frac{\pi}{\alpha}} = \sqrt{2\pi n}, \tag{1.94}$$

we now obtain the Stirling's asymptotic formula

$$n! = e^{n \ln n - n} \sqrt{2\pi n} = \sqrt{2\pi n}(\frac{n}{e})^n. \tag{1.95}$$

From the basic definition, it can be shown that

$$\Gamma(-\frac{1}{2}) = -2\sqrt{\pi}, \quad \Gamma(\frac{1}{2}) = \sqrt{\pi}, \quad \Gamma(\frac{3}{2}) = \frac{\sqrt{\pi}}{2}. \tag{1.96}$$

The standard gamma function can be decomposed into two incomplete functions: the lower incomplete gamma function

$\gamma(\alpha, x)$ and the upper incomplete gamma function $\Gamma(\alpha, x)$ so that $\Gamma(x) = \gamma(\alpha, x) + \Gamma(\alpha, x)$.

The lower incomplete gamma function is defined by

$$\gamma(\alpha, x) = \int_0^x t^{\alpha-1} e^{-t} dt, \tag{1.97}$$

while the upper incomplete gamma function is defined by

$$\Gamma(\alpha, x) = \int_x^\infty t^{\alpha-1} e^{-t} dt. \tag{1.98}$$

Obviously, $\gamma(\alpha, x) \to \Gamma(\alpha)$ as $x \to \infty$. As $\Gamma(\frac{1}{2}) = \sqrt{\pi}$, we have

$$\text{erf}(x) = \frac{1}{\sqrt{\pi}} \gamma(\frac{1}{2}, x^2). \tag{1.99}$$

Another related function is a beta function

$$B(x, y) = \int_0^1 t^{x-1} (1-t)^{y-1} dt. \tag{1.100}$$

From the definition, we know that the beta function is symmetric, $B(x, y) = B(y, x)$. The beta function is linked to $\Gamma$ function by

$$B(x, y) = \frac{\Gamma(x)\Gamma(y)}{\Gamma(x+y)}. \tag{1.101}$$

### 1.5.5 Bessel Functions

Bessel functions come from the solution of the Bessel's equation

$$x^2 y'' + xy' + (x^2 - \lambda^2) y = 0, \tag{1.102}$$

which arises from heat conduction and diffusion problems as well as wave propagation problems. The solution (see later chapters in this book) can be expressed as Taylor's series, and the Bessel function associated with this equation can be defined by

$$J_\lambda(x) = \sum_{n=0}^{\infty} \frac{(-1)^n}{n!\Gamma(n+\lambda+1)} (\frac{x}{2})^{2n+\lambda}, \tag{1.103}$$

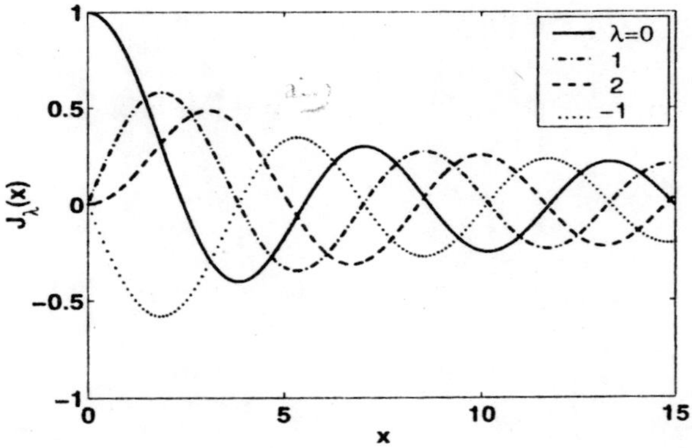


Figure 1.7: Bessel functions.

where $\lambda$ is a real parameter. These are the Bessel functions of the first kind. It can also be defined by the Bessel integral

$$J_\lambda(x) = \frac{1}{2\pi}\int_0^{2\pi} \cos[\lambda t - x\sin t]dt. \tag{1.104}$$

The Bessel functions of the second kind are related to $J_\lambda$, and can be defined by

$$Y_\lambda = \frac{J_\lambda \cos(\lambda\pi) - J_{-\lambda}}{\sin(\lambda\pi)}. \tag{1.105}$$

When $\lambda = k$ is an integer, they have the following properites

$$J_{-k}(x) = (-1)^k J_k(x), \qquad Y_{-k}(x) = (-1)^k Y_k(x). \tag{1.106}$$

The Bessel functions of the first kind are plotted in Figure 1.7.

With these fundamentals of preliminary mathematics, we are now ready to study a wide range of mathematical methods in engineering.

# Chapter 2

# Vector Analysis

Many quantities such as force, velocity, and deformation in engineering and sciences are vectors which have both a magnitude and a direction. The manipulation of vectors is often associated with matrices. In this chapter, we will introduce the basics of vectors and vector analysis.

## 2.1 Vectors

A vector $\mathbf{x}$ is a set of ordered numbers $\mathbf{x} = (x_1, x_2, ..., x_n)$, where its components $x_1, x_2, ..., x_n$ are real numbers. All these vectors form a $n$-dimensional vector space $\mathcal{V}^n$. To add two vectors $\mathbf{x} = (x_1, x_2, ..., x_n)$ and $\mathbf{y} = (y_1, y_2, ..., y_n)$, we simply add their corresponding components.

$$\mathbf{z} = \mathbf{x} + \mathbf{y} = (x_1 + y_1, x_2 + y_2, ..., x_n + y_n), \tag{2.1}$$

and the sum is also a vector. This follows the vector addition parallelogram as shown in Fig 2.1

The addition of vectors has commutability $(\mathbf{u} + \mathbf{v} = \mathbf{v} + \mathbf{u})$ and associativity $[(\mathbf{a} + \mathbf{b}) + \mathbf{c} = \mathbf{a} + (\mathbf{b} + \mathbf{c})]$. Zero vector $\mathbf{0}$ is a special vector that all its components are zeros. The multiplication of a vector $\mathbf{x}$ with a scalar or constant $\alpha$ is carried out by the multiplication of each component,

$$\alpha\mathbf{y} = (\alpha y_1, \alpha y_2, ..., \alpha y_n). \tag{2.2}$$

Thus, $-\mathbf{y} = (-y_1, -y_2, ..., -y_n)$. In addition, $(\alpha\beta)\mathbf{y} = \alpha(\beta\mathbf{y})$ and $(\alpha + \beta)\mathbf{y} = \alpha\mathbf{y} + \beta\mathbf{y}$.

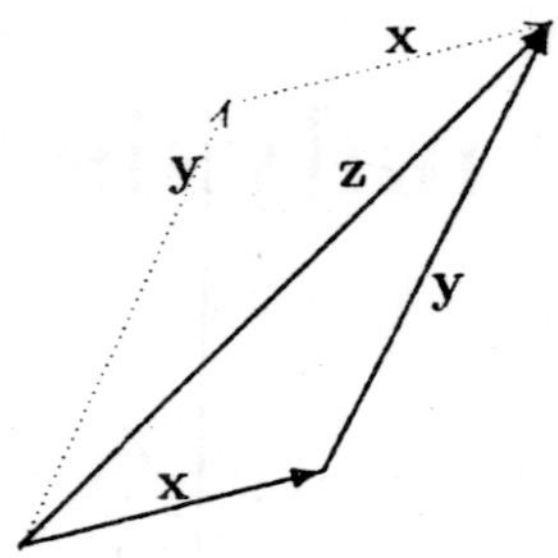


Figure 2.1: Vector addition.

Two nonzero vectors $\mathbf{a}$ and $\mathbf{b}$ are said to be linearly independent if $\alpha\mathbf{a} + \beta\mathbf{b} = 0$ implies that $\alpha = \beta = 0$. If $\alpha, \beta$ are not all zeros, then these two vectors are linearly dependent. Two linearly dependent vectors are parallel $(\mathbf{a}/\!/\mathbf{b})$ to each other. Three linearly dependent vectors $\mathbf{a}, \mathbf{b}, \mathbf{c}$ are in the same plane.

### 2.1.1 Dot Product and Norm

The dot product or inner product of two vectors $\mathbf{x}$ and $\mathbf{y}$ is defined as

$$\mathbf{x} \cdot \mathbf{y} = x_1 y_1 + x_2 y_2 + ... + x_n y_n = \sum_{i=1}^{n} x_i y_i, \tag{2.3}$$

which is a real number. The length or norm of a vector $\mathbf{x}$ is the root of the dot product of the vector itself,

$$|\mathbf{x}| = \|\mathbf{x}\| = \sqrt{\mathbf{x} \cdot \mathbf{x}} = \sqrt{\sum_{i=1}^{n} x_i^2}. \tag{2.4}$$

When $||\mathbf{x}|| = 1$, then it is a unit vector. It is straightforward to check that the dot product has the following properties:

$$\mathbf{x} \cdot \mathbf{y} = \mathbf{y} \cdot \mathbf{x}, \qquad \mathbf{x} \cdot (\mathbf{y} + \mathbf{z}) = \mathbf{x} \cdot \mathbf{y} + \mathbf{x} \cdot \mathbf{z}, \tag{2.5}$$

and

$$(\alpha \mathbf{x}) \cdot (\beta \mathbf{y}) = (\alpha\beta)\mathbf{x} \cdot \mathbf{y}, \tag{2.6}$$

where $\alpha, \beta$ are constants.

If $\theta$ is the angle between two vectors $\mathbf{a}$ and $\mathbf{b}$, then the dot product can also be written

$$\mathbf{a} \cdot \mathbf{b} = ||\mathbf{a}||\ ||\mathbf{b}|| \cos(\theta), \qquad 0 \le \theta \le \pi. \tag{2.7}$$

If the dot product of these two vectors is zero or $\cos(\theta) = 0$ (i.e., $\theta = \pi/2$), then we say that these two vectors are orthogonal.

Rearranging equation (2.7), we obtain a formula to calculate the angle $\theta$ between two vectors

$$\cos(\theta) = \frac{\mathbf{a} \cdot \mathbf{b}}{||\mathbf{a}||\ ||\mathbf{b}||}. \tag{2.8}$$

Since $\cos(\theta) \le 1$, then we get the useful Cauchy-Schwartz inequality:

$$||\mathbf{a} \cdot \mathbf{b}|| \le ||\mathbf{a}||\ ||\mathbf{b}||. \tag{2.9}$$

Any vector $\mathbf{a}$ in a $n$-dimensional vector space $\mathcal{V}^n$ can be written as a combination of a set of $n$ independent basis vectors or orthogonal spanning vectors $\mathbf{e_1}, \mathbf{e_2}, ..., \mathbf{e_n}$, so that

$$\mathbf{a} = \alpha_1 \mathbf{e}_1 + \alpha_2 \mathbf{e}_2 + ... + \alpha_n \mathbf{e}_n = \sum_{i=1}^{n} \alpha_i \mathbf{e}_i, \tag{2.10}$$

where the coefficients/scalars $\alpha_1, \alpha_2, ..., \alpha_n$ are the components of $\mathbf{a}$ relative to the basis $\mathbf{e_1}, \mathbf{e_2} ..., \mathbf{e}_n$. The most common basis vectors are the orthogonal unit vectors. In a three-dimensional case, they are $\mathbf{i} = (1, 0, 0)$, $\mathbf{j} = (0, 1, 0)$, $\mathbf{k} = (0, 0, 1)$ for three $x-, y-, z-$axis, and thus $\mathbf{x} = x_1\mathbf{i} + x_2\mathbf{j} + x_3\mathbf{k}$. The three unit vectors satisfy $\mathbf{i} \cdot \mathbf{j} = \mathbf{j} \cdot \mathbf{k} = \mathbf{k} \cdot \mathbf{i} = 0$.

### 2.1.2 Cross Product

The dot product of two vectors is a scalar or a number. On the other hand, the cross product or outer product of two vectors is a new vector

$$\mathbf{c} = \mathbf{a} \times \mathbf{b}$$

$$= (x_2 y_3 - x_3 y_2)\mathbf{i} + (x_3 y_1 - x_1 y_3)\mathbf{j} + (x_1 y_2 - x_2 y_1)\mathbf{k}, \quad (2.11)$$

which is usually written as

$$\mathbf{a} \times \mathbf{b} = \begin{vmatrix} \mathbf{i} & \mathbf{j} & \mathbf{k} \\ x_1 & x_2 & x_3 \\ y_1 & y_2 & y_3 \end{vmatrix}$$

$$= \begin{vmatrix} x_2 & x_3 \\ y_2 & y_3 \end{vmatrix} \mathbf{i} + \begin{vmatrix} x_3 & x_1 \\ y_3 & y_1 \end{vmatrix} \mathbf{j} + \begin{vmatrix} x_1 & x_2 \\ y_1 & y_2 \end{vmatrix} \mathbf{k}. \quad (2.12)$$

The angle between **a** and **b** can also be expressed as

$$\sin\theta = \frac{\|\mathbf{a} \times \mathbf{b}\|}{\|\mathbf{a}\| \, \|\mathbf{b}\|}. \quad (2.13)$$

In fact, the norm $\|\mathbf{a} \times \mathbf{b}\|$ is the area of the parallelogram formed by **a** and **b**. The vector $\mathbf{c} = \mathbf{a} \times \mathbf{b}$ is perpendicular to both **a** and **b**, following a right-hand rule. It is straightforward to check that the cross product has the following properties:

$$\mathbf{x}\times\mathbf{y} = -\mathbf{y}\times\mathbf{x}, \quad (\mathbf{x}+\mathbf{y})\times\mathbf{z} = \mathbf{x}\times\mathbf{z} + \mathbf{y}\times\mathbf{z}, \quad (2.14)$$

and

$$(\alpha\mathbf{x})\times(\beta\mathbf{y}) = (\alpha\beta)\mathbf{x}\times\mathbf{y}. \quad (2.15)$$

A very special case is $\mathbf{a}\times\mathbf{a} = \mathbf{0}$. For unit vectors, we have

$$\mathbf{i}\times\mathbf{j} = \mathbf{k}, \qquad \mathbf{j}\times\mathbf{k} = \mathbf{i}, \qquad \mathbf{k}\times\mathbf{i} = \mathbf{j}. \quad (2.16)$$

---

□ ***Example 2.1:*** *For two 3-D vectors* $\mathbf{a} = (1, 1, 0)$ *and* $\mathbf{b} = (2, -1, 0)$, *their dot product is*

$$\mathbf{a} \cdot \mathbf{b} = 1 \times 2 + 1 \times (-1) + 0 = 1.$$

*As their moduli are*

$$||\mathbf{a}|| = \sqrt{1^2 + 1^2 + 0^2} = \sqrt{2}, \quad ||\mathbf{b}|| = \sqrt{2^2 + (-1)^2 + 0} = \sqrt{5},$$

*we can calculate the angle $\theta$ between the two vectors. We have*

$$\cos\theta = \frac{\mathbf{a}\cdot\mathbf{b}}{||\mathbf{a}||||\mathbf{b}||} = \frac{1}{\sqrt{2}\sqrt{5}},$$

*or*

$$\theta = \cos^{-1}\frac{1}{\sqrt{10}} \approx 71.56^\circ.$$

*Their cross product is*

$$\mathbf{v} = \mathbf{a}\times\mathbf{b} = (1\times 0 - 0\times(-1), 0\times 1 - 1\times 0, 1\times(-1) - 2\times 1)$$

$$= (0, 0, -3),$$

*which is a vector pointing in the negative $z$-axis direction. The vector $\mathbf{v}$ is perpendicular to both $\mathbf{a}$ and $\mathbf{b}$ because*

$$\mathbf{a}\cdot\mathbf{v} = 1\times 0 + 1\times 0 + 0\times(-3) = 0,$$

*and*

$$\mathbf{b}\cdot\mathbf{v} = 2\times 0 + (-1)\times 0 + 0\times(-3) = 0.$$

□

### 2.1.3 Vector Triple

For two vectors, their product can be either a scalar (dot product) or a vector (cross product). Similarly, the product of triple vectors $\mathbf{a}, \mathbf{b}, \mathbf{c}$ can be either a scalar

$$\mathbf{a}\cdot(\mathbf{b}\times\mathbf{c}) = \begin{vmatrix} a_x & a_y & a_z \\ b_x & b_y & b_z \\ c_x & c_y & c_z \end{vmatrix}, \tag{2.17}$$

or a vector

$$\mathbf{a}\times(\mathbf{b}\times\mathbf{c}) = (\mathbf{a}\cdot\mathbf{c})\mathbf{b} - (\mathbf{a}\cdot\mathbf{b})\mathbf{c}. \tag{2.18}$$

As the dot product of two vectors is the area of a parallelogram, the scalar triple product is the volume of the parallelepiped formed by the three vectors. From the definitions, it is straightforward to prove that

$$\mathbf{a} \cdot (\mathbf{b} \times \mathbf{c}) = \mathbf{b} \cdot (\mathbf{c} \times \mathbf{a}) = \mathbf{c} \cdot (\mathbf{a} \times \mathbf{b}) = -\mathbf{a} \cdot (\mathbf{c} \times \mathbf{b}), \quad (2.19)$$

$$\mathbf{a} \times (\mathbf{b} \times \mathbf{c}) \neq (\mathbf{a} \times \mathbf{b}) \times \mathbf{c}, \quad (2.20)$$

and

$$(\mathbf{a} \times \mathbf{b}) \cdot (\mathbf{c} \times \mathbf{d}) = (\mathbf{a} \cdot \mathbf{c})(\mathbf{b} \cdot \mathbf{d}) - (\mathbf{a} \cdot \mathbf{d})(\mathbf{b} \cdot \mathbf{c}). \quad (2.21)$$

## 2.2 Vector Algebra

### 2.2.1 Differentiation of Vectors

The differentiation of a vector is carried out over each component and treating each component as the usual differentiation of a scalar. Thus, for a position vector

$$\mathbf{P}(t) = x(t)\mathbf{i} + y(t)\mathbf{j} + z(t)\mathbf{k}, \quad (2.22)$$

we can write its velocity as

$$\mathbf{v} = \frac{d\mathbf{P}}{dt} = \dot{x}(t)\mathbf{i} + \dot{y}(t)\mathbf{j} + \dot{z}(t)\mathbf{k}, \quad (2.23)$$

and acceleration as

$$\mathbf{a} = \frac{d^2\mathbf{P}}{dt^2} = \ddot{x}(t)\mathbf{i} + \ddot{y}(t)\mathbf{j} + \ddot{z}(t)\mathbf{k}, \quad (2.24)$$

where $\dot{()} = d()/dt$. Conversely, the integral of $\mathbf{v}$ is

$$\mathbf{P} = \int \mathbf{v} dt + \mathbf{c}, \quad (2.25)$$

where $\mathbf{c}$ is a vector constant.

From the basic definition of differentiation, it is easy to check that the differentiation of vectors has the following properties:

$$\frac{d(\alpha \mathbf{a})}{dt} = \alpha \frac{d\mathbf{a}}{dt}, \quad \frac{d(\mathbf{a} \cdot \mathbf{b})}{dt} = \frac{d\mathbf{a}}{dt} \cdot \mathbf{b} + \mathbf{a} \cdot \frac{d\mathbf{b}}{dt}, \tag{2.26}$$

and

$$\frac{d(\mathbf{a} \times \mathbf{b})}{dt} = \frac{d\mathbf{a}}{dt} \times \mathbf{b} + \mathbf{a} \times \frac{d\mathbf{b}}{dt}. \tag{2.27}$$

### 2.2.2 Kinematics

As an application of vector algebra, let us study the motion along a curved path. In mechanics, there are three coordinate systems which can be used to describe the motion uniquely. The first one is the Cartesian coordinates $(x, y)$ with two unit vectors $\boldsymbol{i}$ (along positive $x$-axis) and $\boldsymbol{j}$ (along positive $y$-axis), and the second one is the polar coordinates $(r, \theta)$ with two unit vectors $\boldsymbol{e}_r$ and $\boldsymbol{e}_\theta$ as shown in Figure 2.2.

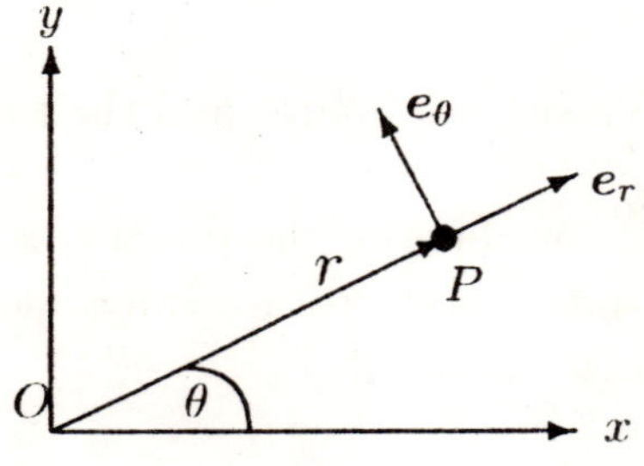


Figure 2.2: Polar coordinates, their unit vectors and their relationship with Cartesian coordinates.

The position vector $\mathbf{r} = x(t)\mathbf{i} + y(t)\mathbf{j}$ at point $P$ at any instance $t$ in the Cartesian coordinates can be expressed as $(r, \theta)$. The velocity vector is

$$\mathbf{v} = \dot{r}\boldsymbol{e}_r + r\dot{\theta}\boldsymbol{e}_\theta. \tag{2.28}$$

and the acceleration is

$$\mathbf{a} = \dot{\mathbf{v}} = (\ddot{r} - r\dot{\theta}^2)\boldsymbol{e}_r + (r\ddot{\theta} + 2\dot{r}\dot{\theta})\boldsymbol{e}_\theta. \tag{2.29}$$

The third coordinate system is the intrinsic coordinate system $(s, \psi)$ where $s$ is the arc length from a reference point (say, point O) and $\psi$ is the angle of the tangent at the point $P$ (see Figure 2.3). The two unit vectors for this systems are $\boldsymbol{e}_t$ along the tangent direction and $\boldsymbol{e}_n$ which is the unit normal of the curve.

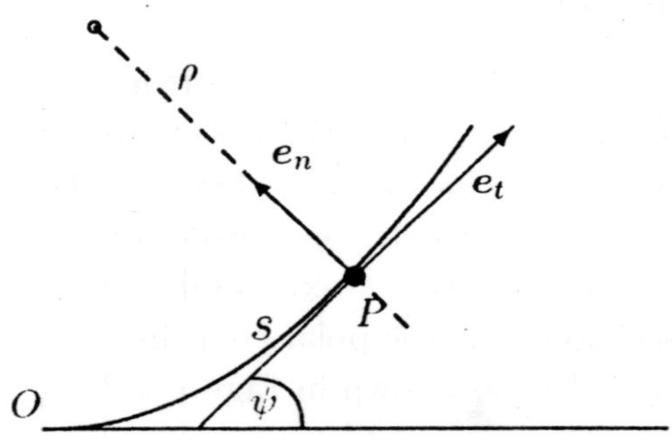


Figure 2.3: Intrinsic coordinates and their unit vectors.

In the intrinsic coordinates, the position is uniquely determined by $(s, \psi)$, and the velocity is always along the tangent. Naturally, the velocity is simply

$$\mathbf{v} = \dot{s}\boldsymbol{e}_t. \tag{2.30}$$

The acceleration becomes

$$\mathbf{a} = \ddot{s}\boldsymbol{e}_t + \frac{\dot{s}^2}{\rho}\boldsymbol{e}_n, \tag{2.31}$$

where $\rho$ is the radius of the curvature at point $P$.

For the circular motion such as a moving bicycle wheel as shown in Figure 2.4, the three coordinate systems are interconnected. In a rotating reference frame with an angular velocity

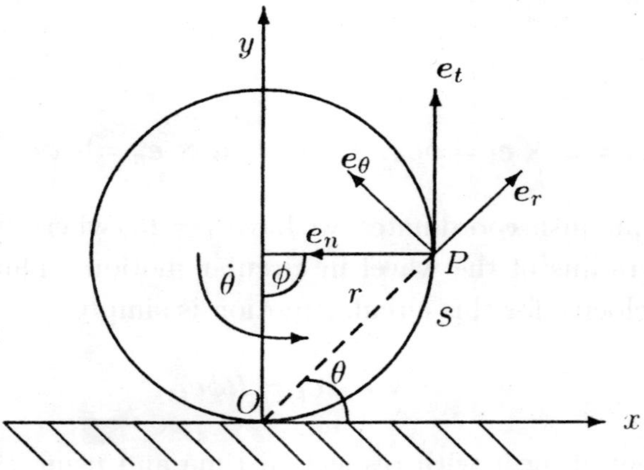


Figure 2.4: Three coordinate systems for a wheel in circular motion.

$\omega = \dot{\theta}\mathbf{k}$ where $\mathbf{k}$ point to the $z$-axis, the velocity and acceleration at any point (say) $P$ can be calculated using another fixed point $A$ on the rotating body (or wheel). The velocity is

$$\mathbf{v}_P = \mathbf{v}_A + \frac{d\mathbf{r}}{dt}\Big|_A + \boldsymbol{\omega} \times \mathbf{r}, \tag{2.32}$$

and the acceleration is

$$\mathbf{a}_P = \mathbf{a}_A + \frac{d^2\mathbf{r}}{dt^2}\Big|_A + \frac{d\omega}{dt} \times \mathbf{r} + \mathbf{a}_{Cor} + \mathbf{a}_{Cent}, \tag{2.33}$$

where

$$\mathbf{a}_{Cor} = 2\omega \times \frac{d\mathbf{r}}{dt}\Big|_A, \tag{2.34}$$

is the Coriolis acceleration, and

$$\mathbf{a}_{Cent} = \boldsymbol{\omega} \times (\boldsymbol{\omega} \times \mathbf{r}), \tag{2.35}$$

is the centripetal acceleration. It is worth noting that the velocity $\mathbf{v}_A$ and acceleration $\mathbf{a}_A$ is the velocity and acceleration in a non-rotating frame or an inertia frame.

In addition, the differentiation of the unit vectors are connected by

$$\dot{\boldsymbol{e}}_r = \boldsymbol{\omega} \times \boldsymbol{e}_r = \dot{\theta}\boldsymbol{e}_\theta, \qquad \dot{\boldsymbol{e}}_\theta = \boldsymbol{\omega} \times \boldsymbol{e}_\theta = -\dot{\theta}\boldsymbol{e}_r. \tag{2.36}$$

and

$$\dot{\boldsymbol{e}}_t = \boldsymbol{\omega} \times \boldsymbol{e}_t = \dot{\phi}\boldsymbol{e}_n, \qquad \dot{\boldsymbol{e}}_n = \boldsymbol{\omega} \times \boldsymbol{e}_n = -\dot{\phi}\boldsymbol{e}_t. \tag{2.37}$$

In the intrinsic coordinates, we have $s = R\phi$ where $R$ =constant is the radius of the wheel in circular motion. Thus, $\dot{s} = R\dot{\phi}$. The velocity for this circular motion is simply

$$\mathbf{v} = \dot{s}\boldsymbol{e}_t = R\dot{\phi}\boldsymbol{e}_t. \tag{2.38}$$

Differentiating it with respect to time and using the relationships of unit vectors, we have

$$\mathbf{a} = \dot{\mathbf{v}} = R\ddot{\phi}\boldsymbol{e}_t + R\dot{\phi}^2\boldsymbol{e}_n, \tag{2.39}$$

where the unit vectors are

$$\boldsymbol{e}_t = \cos\phi\mathbf{i} + \sin\phi\mathbf{j}, \qquad \boldsymbol{e}_n = -\sin\phi\mathbf{i} + \cos\phi\mathbf{j}. \tag{2.40}$$

---

□ ***Example 2.2:*** *A car is travelling rapidly along a curved path with a speed of 30 m/s at a given instance. The car is fitted with an accelerometer and it shows that the car is accelerating along the curved path at 2 m/s$^2$. The accelerometer also indicates that the component of the acceleration perpendicular to the travelling direction is 5 m/s$^2$. What is the direction of the total acceleration at this instance? What is the radius of the curvature? Suppose the car has a height of 2 meters and a width of 1.6 meters, is there any danger of toppling over?*

*Let $\theta$ be the angle between the acceleration vector and the velocity vector, and let $a$ be the magnitude of the total acceleration. In the intrinsic coordinates, the velocity is $\mathbf{v} = \dot{s}\boldsymbol{e}_t = 30\boldsymbol{e}_t$. The acceleration is given by*

$$\mathbf{a} = \ddot{s}\boldsymbol{e}_t + \frac{\dot{s}^2}{\rho}\boldsymbol{e}_n = a(\cos\theta\boldsymbol{e}_t + \sin\theta\boldsymbol{e}_n).$$

*Therefore, we have*

$$\frac{\dot{s}^2}{\rho} = \frac{30^2}{\rho} = a\sin\theta = 5,$$

*or the instantaneous radius of curvature is $\rho = 30^2/5 = 180$m. We know that the magnitude of the acceleration is $a = \sqrt{2^2+5^2} = \sqrt{29}$. The angle is*

$$\theta = \tan^{-1}\frac{5}{2} \approx 68.20°.$$

*In addition, we can assume that the centre of gravity is approximately at its geometrical centre. Thus, the centre is 1m above the road surface and 0.8m from the edges of the outer wheels. If we take the moment about the axis through the two contact points of the outer wheels, we have the total moment*

$$1 \times M\frac{v^2}{\rho} - 0.8Mg \approx -2.8M < 0,$$

*where $M$ is the mass of the car. There is no danger of toppling over. However, if the car speeds up to $v = 42$ m/s (about 95 miles per hour), there is a danger of toppling over when the moment of the weight is just balanced by the moment of the centripetal force.* □

### 2.2.3 Line Integral

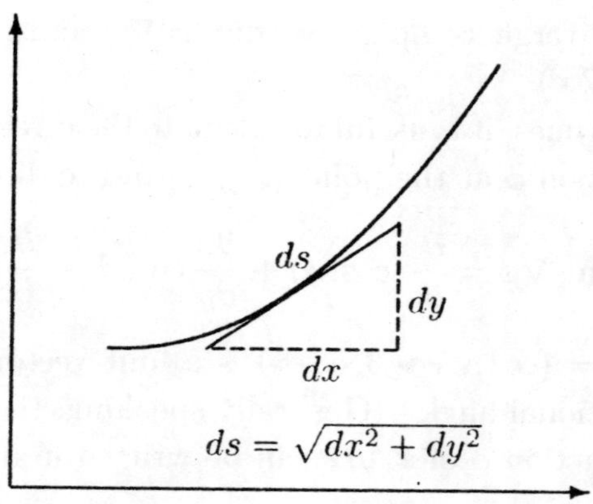


Figure 2.5: Arc length along a curve.

An important class of integrals in this context is the line integral which integrates along a curve $\mathbf{r}(x,y,z) = x\mathbf{i}+y\mathbf{j}+z\mathbf{k}$. For example, in order to calculate the arc length $L$ of curve $\mathbf{r}$

as shown in Figure 2.5, we have to use the line integral.

$$L = \int_{s_0}^{s} ds = \int_{s_0}^{s} \sqrt{dx^2 + dy^2} = \int_{x_0}^{x} \sqrt{1 + (\frac{dy}{dx})^2} dx. \quad (2.41)$$

---

□ ***Example 2.3:*** *The arc length of the parabola $y(x) = \frac{1}{2}x^2$ from $x = 0$ to $x = 1$ is given by*

$$L = \int_0^1 \sqrt{1 + y'^2} dx = \int_0^1 \sqrt{1 + x^2} dx$$

$$= \frac{1}{2}[x\sqrt{1 + x^2} + \ln(x + \sqrt{1 + x^2})]\Big|_0^1$$

$$= \frac{1}{2}[\sqrt{2} - \ln(\sqrt{2} - 1)] \approx 1.14779.$$

□

---

### 2.2.4 Three Basic Operators

Three important operators commonly used in vector analysis, especially in fluid dynamics, are the gradient operator (grad or $\nabla$), the divergence operator (div or $\nabla\cdot$) and the curl operator (curl or $\nabla\times$).

Sometimes, it is useful to calculate the directional derivative of a function $\phi$ at the point $(x, y, z)$ in the direction of $\mathbf{n}$

$$\frac{\partial \phi}{\partial \mathbf{n}} = \mathbf{n} \cdot \nabla \phi = \frac{\partial \phi}{\partial x} \cos(\alpha) + \frac{\partial \phi}{\partial y} \cos(\beta) + \frac{\partial \phi}{\partial z} \cos(\gamma), \quad (2.42)$$

where $\mathbf{n} = (\cos\alpha, \cos\beta, \cos\gamma)$ is a unit vector and $\alpha, \beta, \gamma$ are the directional angles. Generally speaking, the gradient of any scalar function $\phi$ of $x, y, z$ can be written in a similar way,

$$\text{grad}\phi = \nabla\phi = \frac{\partial \phi}{\partial x}\mathbf{i} + \frac{\partial \phi}{\partial y}\mathbf{j} + \frac{\partial \phi}{\partial z}\mathbf{k}. \quad (2.43)$$

This is equivalent to applying the del operator $\nabla$ to the scalar function $\phi$

$$\nabla = \frac{\partial}{\partial x}\mathbf{i} + \frac{\partial}{\partial y}\mathbf{j} + \frac{\partial}{\partial z}\mathbf{k}. \quad (2.44)$$

The direction of the gradient operator on a scalar field gives a vector field. The gradient operator has the following properties:

$$\nabla(\alpha\psi+\beta\phi) = \alpha\nabla\psi+\beta\nabla\phi, \qquad \nabla(\psi\phi) = \psi\nabla\phi+\phi\nabla\psi, \quad (2.45)$$

where $\alpha, \beta$ are constants and $\psi, \phi$ are scalar functions.

For a vector field

$$\mathbf{u}(x, y, z) = u_1(x, y, z)\mathbf{i} + u_2(x, y, z)\mathbf{j} + u_3(x, y, z)\mathbf{k}, \quad (2.46)$$

the application of the operator $\nabla$ can lead to either a scalar field or a vector field, depending on how the del operator applies to the vector field. The divergence of a vector field is the dot product of the del operator $\nabla$ and $\mathbf{u}$

$$\text{div } \mathbf{u} = \nabla \cdot \mathbf{u} = \frac{\partial u_1}{\partial x} + \frac{\partial u_2}{\partial y} + \frac{\partial u_3}{\partial z}, \quad (2.47)$$

and the curl of $\mathbf{u}$ is the cross product of the del operator and the vector field $\mathbf{u}$

$$\text{curl } \mathbf{u} = \nabla \times \mathbf{u} = \begin{vmatrix} \mathbf{i} & \mathbf{j} & \mathbf{k} \\ \frac{\partial}{\partial x} & \frac{\partial}{\partial y} & \frac{\partial}{\partial z} \\ u_1 & u_2 & u_3 \end{vmatrix}. \quad (2.48)$$

It is straightforward to verify the following useful identities associated with the $\nabla$ operator:

$$\nabla \cdot \nabla \times \mathbf{u} = 0. \quad (2.49)$$

$$\nabla \times \nabla\psi = 0. \quad (2.50)$$

$$\nabla \times (\psi\mathbf{u}) = \psi\nabla \times \mathbf{u} + (\nabla\psi) \times \mathbf{u}. \quad (2.51)$$

$$\nabla \cdot (\psi\mathbf{u}) = \psi\nabla \cdot \mathbf{u} + (\nabla\psi) \cdot \mathbf{u}. \quad (2.52)$$

$$\nabla \times (\nabla \times \mathbf{u}) = \nabla(\nabla \cdot \mathbf{u}) - \nabla^2\mathbf{u}. \quad (2.53)$$

One of the most common operators in engineering and science is the Laplacian operator is

$$\nabla^2\Psi = \nabla \cdot (\nabla\Psi) = \frac{\partial^2\Psi}{\partial x^2} + \frac{\partial^2\Psi}{\partial y^2} + \frac{\partial^2\Psi}{\partial z^2}. \quad (2.54)$$

for Laplace's equation

$$\Delta\Psi = \nabla^2\Psi = 0. \tag{2.55}$$

In engineering mathematics, it is sometimes necessary to express the Laplace equation in other coordinates. In cylindrical polar coordinates $(r, \phi, z)$, we have

$$\nabla\cdot\mathbf{u} = \frac{1}{r}\frac{\partial(ru_r)}{\partial r} + \frac{1}{r}\frac{\partial u_\phi}{\partial\phi} + \frac{\partial u_z}{\partial z}. \tag{2.56}$$

The Laplace equation becomes

$$\nabla^2\Psi = \frac{\partial^2\Psi}{\partial r^2} + \frac{1}{r}\frac{\partial\Psi}{\partial r} + \frac{1}{r^2}\frac{\partial^2\Psi}{\partial\phi^2} + \frac{\partial^2\Psi}{\partial z^2}. \tag{2.57}$$

In spherical polar coordinates $(r, \theta, \phi)$, we have

$$\nabla\cdot\mathbf{u} = \frac{1}{r^2}\frac{\partial^2(r^2u_r)}{\partial r^2} + \frac{1}{r\sin\theta}\frac{\partial(\sin\theta u_\theta)}{\partial\theta} + \frac{1}{r\sin\theta}\frac{\partial u_\phi}{\partial\phi}. \tag{2.58}$$

The Laplace equation can be written as

$$\nabla^2\Psi = \frac{1}{r^2}\frac{\partial}{\partial r}[r^2\frac{\partial\Psi}{\partial r}]$$

$$+\frac{1}{r^2\sin\theta}\frac{\partial}{\partial\theta}[\sin\theta\frac{\partial\Psi}{\partial\theta}] + \frac{1}{r^2\sin^2\theta}\frac{\partial^2\Psi}{\partial\phi^2}. \tag{2.59}$$

### 2.2.5 Some Important Theorems

The Green theorem is an important theorem, especially in fluid dynamics and the finite element analysis. For a vector field $\mathbf{Q} = u\mathbf{i}+v\mathbf{j}$ in a 2-D region $\Omega$ with the boundary $\Gamma$ and the unit outer normal $\mathbf{n}$ and unit tangent $\mathbf{t}$. The theorems connect the integrals of divergence and curl with other integrals. Gauss's theorem states:

$$\iiint_\Omega(\nabla\cdot\mathbf{Q})d\Omega = \iint_S \mathbf{Q}\cdot\mathbf{n}dS, \tag{2.60}$$

which connects the volume integral to the surface integral.

Another important theorem is Stokes's theorem:

$$\iint_S (\nabla \times \mathbf{Q}) \cdot \mathbf{k} d\mathbf{S} = \oint_\Gamma \mathbf{Q} \cdot \mathbf{t} d\Gamma = \oint_\Gamma \mathbf{Q} \cdot d\mathbf{r}, \tag{2.61}$$

which connects the surface integral to the corresponding line integral.

In our simple 2-D case, this becomes

$$\oint (udx + vdy) = \iint_\Omega (\frac{\partial v}{\partial x} - \frac{\partial u}{\partial y}) dxdy. \tag{2.62}$$

For any scalar functions $\psi$ and $\phi$, the useful Green's first identity can be written as

$$\oint_{\partial\Omega} \psi \nabla \phi d\Gamma = \int_\Omega (\psi \nabla^2 \phi + \nabla \psi \cdot \nabla \phi) d\Omega, \tag{2.63}$$

where $d\Omega = dxdydz$. By using this identity twice, we get Green's second identity

$$\oint_{\partial\Omega} (\psi \nabla \phi - \phi \nabla \psi) d\Gamma = \int_\Omega (\psi \nabla^2 \phi - \phi \nabla^2 \psi) d\Omega. \tag{2.64}$$

## 2.3 Applications

In order to show the wide applications of vector analysis, let us apply them to study the mechanical and flow problems.

### 2.3.1 Conservation of Mass

The mass conservation in flow mechanics can be expressed in either integral form (weak form) or differential form (strong form). For any enclosed volume $\Omega$, the total mass which leaves or enters the surface $S$ is

$$\oint_S \rho \boldsymbol{u} \cdot d\mathbf{A},$$

where $\rho(x, y, z, t)$ and $\boldsymbol{u}(x, y, z, t)$ are the density and the velocity of the fluid, respectively. The rate of change of mass in $\Omega$ is

$$\frac{\partial}{\partial t} \int \rho dV.$$

The mass conservation requires that the rate of loss of mass through the surface $S$ is balanced by the rate of change in $\Omega$. Therefore, we have

$$\oint_S \rho \boldsymbol{u} \cdot d\mathbf{A} + \frac{\partial}{\partial t}\int dV = 0.$$

Using Gauss's theorem for the surface integral, we have

$$\int_\Omega \nabla \cdot (\rho \boldsymbol{u}) dV + \frac{\partial}{\partial t}\int_\Omega \rho dV = 0.$$

Interchange of the integration and differentiation in the second term, we have

$$\int_\Omega [\frac{\partial \rho}{\partial t} + \nabla \cdot (\rho \boldsymbol{u})] dV = 0.$$

This is the integral form or weak form of the conservation of mass. This is true for any volume at any instance, and subsequently the only way that this is true for all possible choice of $\Omega$ is

$$\frac{\partial \rho}{\partial t} + \nabla \cdot (\rho \boldsymbol{u}) = 0,$$

which is the differential form or strong form of mass conservation. The integral form is more useful in numerical methods such as finite volume methods and finite element methods, while the differential form is more natural for mathematical analysis.

### 2.3.2 Saturn's Rings

We all know that Saturn's ring system ranks among the most spectacular phenomena in the solar system. The ring system has a diameter of 270,000 km, yet its thickness does not exceed 100 meters. The sizes of particles in the rings vary from centimeters to several meters, and this size distribution is consistent with the distribution caused by repeated collision. The ring system has very complicated structures. One natural question is why the formed structure is a ring system, why not a spherical shell system? This is a challenging topic which has not yet fully understood. However, under some reasonable assumptions, we can understand why the ring system is so.

When the debris particles surrounding a planet will ultimately settle into flat circular rings or disks, which are the natural consequence of energy dissipation in rotating systems. The interparticle collisions dissipate energy while conserving the total angular momentum. Laplace in 1802 showed that such rings could not be solid because the tensile strength of the known materials was too small to resist tidal forces from Saturn. Later, Maxwell in 1890 showed that a fluid or gaseous ring was unstable, therefore, the rings must be particulate.

Suppose the whole particulate system consists of $N$ particles ($i = 1, 2, ..., N$). Its total angular momentum is $h$. By choosing a coordinate system so that $(x, y)$ plane coincides with the plane of the rings, and the $z$-axis (along $\boldsymbol{k}$ direction) is normal to this plane. If we now decompose the velocity of each particle into $\boldsymbol{v}_i = (v_{ir}, v_{i\theta}, v_{iz})$, the total angular momentum is then

$$h = \boldsymbol{k} \cdot [\sum_{i=1}^{N} m_i \boldsymbol{r}_i \times \boldsymbol{v}_i]$$

$$= \sum_{i=1}^{N} m_i(\boldsymbol{r}_i \times \boldsymbol{v}_{iz}) \cdot \boldsymbol{k} + \sum_{i=1}^{N} m_i(\boldsymbol{r}_i \times \boldsymbol{v}_{ir}) \cdot \boldsymbol{k} + \sum_{i=1}^{N} m_i(\boldsymbol{r}_i \times \boldsymbol{v}_{i\theta}) \cdot \boldsymbol{k}. \tag{2.65}$$

The first two terms disappear because $\boldsymbol{v}_{iz}$ is parallel to $\boldsymbol{k}$ and axial velocity does not contribute to the angular momentum. So only the tangential terms are non-zero, and we have

$$h = \sum_{i=1}^{N} m_i r_i v_{i\theta}. \tag{2.66}$$

The total mechanical energy is

$$E = \frac{1}{2}\sum_{i=1}^{N} m_i(v_{ir}^2 + v_{i\theta}^2 + v_{iz}^2) + \sum_{i=1}^{N} m_i\psi(r_i)_{Saturn}, \tag{2.67}$$

where $\psi(r_i)$ is the potential per unit mass due to Saturn's gravity. The interparticle collisions will dissipate the energy, therefore, the system will evolve towards an energy minimum. From both expressions for $h$ and $E$, we can see that the minimization of $E$ while $h$ is held constant requires that $v_{iz} \to 0$ and $v_{i\theta} \to 0$. This means that the collisions dissipate energy while flattening the system into a disk or rings.

Now let us see why the minimization of the rotational energy will also lead to the same conclusion of ring formation. Loosely speaking, we can assume that the angular velocity $\omega = \dot{\theta}$ is almost the same for all particles as $t \to \infty$ (or any reasonable long time) so that collisions are no longer significant or the rate of energy dissipation is small. If there are different angular velocities, one particle may move faster and ultimately collides with other particles, subsequently redistributing or changing its angular velocity. If we further assume that the potential energy does not change significantly (this is true if the particles do not move significantly along the radial direction), thus the minimization of total energy leads to the minimization of the total rotational energy.

This will essentially lead to a quasi-steady state. With these assumptions, we have $v_{i\theta} = r_i\omega$. Therefore, the angular momentum becomes

$$h = \sum_{i=1} m_i r_i^2 \omega = I\omega, \qquad I = \sum_{i=1}^{N} m_i r_i^2,$$

where $I$ the moment of inertia of the particulate system. The total rotational energy is

$$T = \frac{1}{2}\sum_{i=1}^{N} m_i r_i^2 \omega = \frac{1}{2} I\omega^2 = \frac{1}{2}\frac{h^2}{I} \to T_{\min}.$$

In order to minimize $T$, we have to maximize $I$ because $h$ is constant. For a disk with a radius $a$, a thickness $t \ll a$ and the total mass $m$, we have

$$I = \int_V r^2 dm = t\int_0^R r^2 \rho r dr \int_0^{2\pi} d\theta = 2\pi t\rho \int_0^R r^3 dr = \pi t\rho \frac{R^4}{2}.$$

Using the density $\rho = m/(t\pi R^2)$, we have

$$I_{\text{disk}} = \frac{1}{2} m R^2.$$

If all the mass is concentrated at a ring, we have

$$I_{\text{ring}} = m R^2.$$

Similarly, for a solid ball with the same mass and same radius $R$, we have

$$I_{\text{ball}} = \frac{2}{5} m R^2.$$

For a spherical shell, we have

$$I_{\text{sphere}} = \frac{2}{3} m R^2.$$

Therefore, we have

$$I_{\text{ring}} > I_{\text{disk}} > I_{\text{sphere}} > I_{\text{ball}}.$$

This means that the total rotational energy is minimized if the particle system evolves into a ring or at least a disk. This is probably the main reason why the planetary system and rings are formed.

# Chapter 3

# Matrix Algebra

## 3.1 Matrix

Matrices are widely used in almost all engineering subjects. A matrix is a table or array of numbers or functions arranged in rows and columns. The elements or entries of a matrix $\mathbf{A}$ are often denoted as $a_{ij}$. A matrix $\mathbf{A}$ has $m$ rows and $n$ columns,

$$\mathbf{A} = [a_{ij}] = \begin{pmatrix} a_{11} & a_{12} & \dots & a_{1j} & \dots & a_{1n} \\ a_{21} & a_{22} & \dots & a_{2j} & \dots & a_{2n} \\ \vdots & \vdots & & a_{ij} & \dots & \vdots \\ a_{m1} & a_{m2} & \dots & a_{mj} & \dots & a_{mn} \end{pmatrix}. \tag{3.1}$$

we say the size of $\mathbf{A}$ is $m$ by $n$, or $m \times n$. $\mathbf{A}$ is square if $m = n$. For example,

$$\mathbf{A} = \begin{pmatrix} 1 & 2 & 3 \\ 4 & 5 & 6 \end{pmatrix}, \qquad \mathbf{B} = \begin{pmatrix} e^x & \sin x \\ -i\cos x & e^{i\theta} \end{pmatrix}, \tag{3.2}$$

and

$$\mathbf{u} = \begin{pmatrix} u \\ v \\ w \end{pmatrix}, \tag{3.3}$$

where $\mathbf{A}$ is a $2\times 3$ matrix, $\mathbf{B}$ is a $2\times 2$ square matrix, and $\mathbf{u}$ is a $3\times 1$ column matrix or column vector.

The sum of two matrices **A** and **B** is only possible if they have the same size $m \times n$, and their sum, which is also $m \times n$, is obtained by adding corresponding entries

$$\mathbf{C} = \mathbf{A} + \mathbf{B}, \qquad c_{ij} = a_{ij} + b_{ij}, \tag{3.4}$$

where $(i = 1, 2, ..., m; j = 1, 2, ..., n)$. We can multiply a matrix **A** by a scalar $\alpha$ by multiplying each entry by $\alpha$. The product of two matrices is only possible if the number of columns of **A** is the same as the number of rows of **B**. That is to say, if **A** is $m \times n$ and **B** is $n \times r$, then the product **C** is $m \times r$,

$$c_{ij} = (AB)_{ij} = \sum_{k=1}^{n} a_{ik} b_{kj}. \tag{3.5}$$

If **A** is a square matrix, then we have $\mathbf{A}^n = \overbrace{AA...A}^{n}$. The multiplications of matrices are generally not commutive, i.e., $\mathbf{AB} \neq \mathbf{BA}$. However, the multiplication has associativity $\mathbf{A}(\mathbf{uv}) = (\mathbf{Au})\mathbf{v}$ and $\mathbf{A}(\mathbf{u} + \mathbf{v}) = \mathbf{Au} + \mathbf{Av}$.

The transpose $\mathbf{A}^T$ of **A** is obtained by switching the position of rows and columns, and thus $\mathbf{A}^T$ will be $n \times m$ if **A** is $m \times n$, $(a^T)_{ij} = a_{ji}, (i = 1, 2, ..., m; j = 1, 2, ..., n)$. In general, we have

$$(\mathbf{A}^T)^T = \mathbf{A}, \qquad (\mathbf{AB})^T = \mathbf{B}^T \mathbf{A}^T. \tag{3.6}$$

The differentiation and integral of a matrix are done on each member element. For example, for a $2 \times 2$ matrix

$$\frac{d\mathbf{A}}{dt} = \dot{\mathbf{A}} = \begin{pmatrix} \frac{da_{11}}{dt} & \frac{da_{12}}{dt} \\ \frac{da_{21}}{dt} & \frac{da_{22}}{dt} \end{pmatrix}, \tag{3.7}$$

and

$$\int \mathbf{A} dt = \begin{pmatrix} \int a_{11} dt & \int a_{12} dt \\ \int a_{21} dt & \int a_{22} dt \end{pmatrix}. \tag{3.8}$$

A diagonal matrix **A** is a square matrix whose every entry off the main diagonal is zero ($a_{ij} = 0$ if $i \neq j$). Its diagonal elements or entries may or may not have zeros. For example, the matrix

$$\mathbf{I} = \begin{pmatrix} 1 & 0 & 0 \\ 0 & 1 & 0 \\ 0 & 0 & 1 \end{pmatrix} \tag{3.9}$$

is a 3×3 identity or unitary matrix. In general, we have

$$\mathbf{AI} = \mathbf{IA} = \mathbf{A}. \tag{3.10}$$

A zero or null matrix $\mathbf{0}$ is a matrix with all of its elements being zero.

## 3.2 Determinant

The determinant of a square matrix $\mathbf{A}$ is a number or scalar obtained by the following recursive formula or the cofactor or Laplace expansion by column or row. For example, expanding by row $k$, we have

$$\det(\mathbf{A}) = |\mathbf{A}| = \sum_{j=1}^{n} (-1)^{k+j} a_{kj} M_{kj}, \tag{3.11}$$

where $M_{ij}$ is the determinant of a minor matrix of $\mathbf{A}$ obtained by deleting row $i$ and column $j$. For a simple 2×2 matrix, its determinant simply becomes

$$\begin{vmatrix} a_{11}, & a_{12} \\ a_{21} & a_{22} \end{vmatrix} = a_{11}a_{22} - a_{12}a_{21}. \tag{3.12}$$

It is easy to verify that the determinant has the following properties:

$$|\alpha\mathbf{A}| = \alpha|\mathbf{A}|, \qquad |\mathbf{A}^T| = |\mathbf{A}|, \qquad |\mathbf{AB}| = |\mathbf{A}||\mathbf{B}|. \tag{3.13}$$

where $\mathbf{A}$ and $\mathbf{B}$ are the same size ($n \times n$).

A $n \times n$ square matrix is singular if $|\mathbf{A}| = 0$, and is nonsingular if and only if $|\mathbf{A}| \neq 0$. The trace of a square matrix $\mathrm{tr}(\mathbf{A})$

is defined as the sum of the diagonal elements,

$$\mathrm{tr}(\mathbf{A}) = \sum_{i=1}^{n} a_{ii} = a_{11} + a_{22} + ... + a_{nn}. \tag{3.14}$$

The rank of a matrix **A** is the number of linearly independent vectors forming the matrix. Generally, the rank of **A** is $\mathrm{rank}(\mathbf{A}) \le \min(m, n)$. For a $n \times n$ square matrix **A**, it is nonsingular if $\mathrm{rank}(\mathbf{A}) = n$.

From the basic definitions, it is straightforward to prove the following

$$(\mathbf{AB...Z})^{\mathbf{T}} = \mathbf{Z}^{\mathbf{T}}...\mathbf{B}^{\mathbf{T}}\mathbf{A}^{\mathbf{T}}, \tag{3.15}$$

$$|\mathbf{AB....Z}| = |\mathbf{A}||\mathbf{B}|...|\mathbf{Z}|, \tag{3.16}$$

$$\mathrm{tr}(\mathbf{A}) = \mathrm{tr}(\mathbf{A}^T), \tag{3.17}$$

$$\mathrm{tr}(\mathbf{A} + \mathbf{B}) = \mathrm{tr}(\mathbf{A}) + tr(\mathbf{B}), \tag{3.18}$$

$$\mathrm{tr}(\mathbf{AB}) = \mathrm{tr}(\mathbf{BA}), \tag{3.19}$$

$$\det(\mathbf{A}^{-1}) = \frac{1}{\det(\mathbf{A})}, \tag{3.20}$$

$$\det(\mathbf{AB}) = \det(\mathbf{A})\det(\mathbf{B}). \tag{3.21}$$

## 3.3 Inverse

The inverse matrix $\mathbf{A}^{-1}$ of a square matrix **A** is defined as

$$\mathbf{A}^{-1}\mathbf{A} = \mathbf{A}\mathbf{A}^{-1} = \mathbf{I}. \tag{3.22}$$

It is worth noting that the unit matrix **I** has the same size as **A**. The inverse of a square matrix exists if and only if **A** is nonsingular or $\det(\mathbf{A}) \neq 0$. From the basic definitions, it is straightforward to prove that the inverse of a matrix has the following properties

$$(\mathbf{A}^{-1})^{-1} = \mathbf{A}, \qquad (\mathbf{A}^{\mathbf{T}})^{-1} = (\mathbf{A}^{-1})^{\mathbf{T}}, \tag{3.23}$$

and

$$(\mathbf{AB})^{-1} = \mathbf{B}^{-1}\mathbf{A}^{-1}. \tag{3.24}$$

A simple useful formula for obtaining the inverse of a $2 \times 2$ matrix is

$$\begin{pmatrix} a & b \\ c & d \end{pmatrix}^{-1} = \frac{1}{(ad-bc)} \begin{pmatrix} d & -b \\ -c & a \end{pmatrix}. \tag{3.25}$$

---

□ ***Example 3.1:*** For two matrices

$$\mathbf{A} = \begin{pmatrix} 1 & 2 & 3 \\ -1 & 1 & 0 \\ 3 & 2 & 2 \end{pmatrix}, \quad \mathbf{B} = \begin{pmatrix} 1 & -1 \\ 2 & 3 \\ 1 & 7 \end{pmatrix},$$

*we have*

$$\mathbf{AB} = \mathbf{V} = \begin{pmatrix} V_{11} & V_{12} \\ V_{21} & V_{22} \\ V_{31} & V_{32} \end{pmatrix},$$

*where*

$$V_{11} = 1 \times 1 + 2 \times 2 + 3 \times 1 = 8, \; V_{12} = 1 \times (-1) + 2 \times 3 + 3 \times 7 = 26;$$

$$V_{21} = -1 \times 1 + 1 \times 2 + 0 \times 1 = 1, \; V_{22} = -1 \times (-1) + 1 \times 3 + 0 \times 7 = 4;$$

$$V_{31} = 3 \times 1 + 2 \times 2 + 2 \times 1 = 9, \; V_{32} = 3 \times (-1) + 2 \times 3 + 2 \times 7 = 17.$$

*Thus,*

$$\mathbf{AB} = \mathbf{V} = \begin{pmatrix} 8 & 26 \\ 1 & 4 \\ 9 & 17 \end{pmatrix}.$$

*However,* **BA** *does not exist. The transpose matrices of* **A** *and* **B** *are*

$$\mathbf{A}^T = \begin{pmatrix} 1 & -1 & 3 \\ 2 & 1 & 2 \\ 3 & 0 & 2 \end{pmatrix}, \quad \mathbf{B}^T = \begin{pmatrix} 1 & 2 & 1 \\ -1 & 3 & 7 \end{pmatrix}.$$

*Similarly, we have*

$$\mathbf{B}^T\mathbf{A}^T = \begin{pmatrix} 8 & 1 & 9 \\ 26 & 4 & 17 \end{pmatrix} = \mathbf{V}^T = (AB)^T.$$

*The inverse of* **A** *is*

$$\mathbf{A}^{-1} = \frac{1}{9}\begin{pmatrix} -2 & -2 & 3 \\ -2 & 7 & 3 \\ 5 & -4 & -3 \end{pmatrix},$$

*and the determinant of* **A** *is*

$$\det |\mathbf{A}| = -9.$$

*The trace of* **A** *is*

$$\text{tr}(\mathbf{A}) = A_{11} + A_{22} + A_{33} = 1 + 1 + 2 = 4.$$

□

## 3.4 Matrix Exponential

Sometimes, we need to calculate exp[**A**], where **A** is a square matrix. In this case, we have to deal with matrix exponentials. The exponential of a square matrix **A** is defined as

$$e^{\mathbf{A}} \equiv \sum_{n=0}^{\infty} \frac{1}{n!}\mathbf{A}^n = \mathbf{I} + \mathbf{A} + \frac{1}{2}\mathbf{A}^2 + .... \tag{3.26}$$

where **I** is a unity matrix with the same size as **A**, and $\mathbf{A}^2 = \mathbf{AA}$ and so on. This (rather odd) definition in fact provides a method to calculate the matrix exponential. The matrix exponentials are very useful in solving systems of differential equations.

---

□ ***Example 3.2:*** *For a simple matrix*

$$\mathbf{A} = \begin{pmatrix} t & 0 \\ 0 & t \end{pmatrix},$$

*we have*

$$e^{\mathbf{A}} = \begin{pmatrix} e^t & 0 \\ 0 & e^t \end{pmatrix}.$$

*For*

$$\mathbf{A} = \begin{pmatrix} t & t \\ t & t \end{pmatrix},$$

we have

$$e^{\mathbf{A}} = \begin{pmatrix} \frac{1}{2}(1+e^{2t}) & \frac{1}{2}(e^{2t}-1) \\ \frac{1}{2}(e^{2t}-1) & \frac{1}{2}(1+e^{2t}) \end{pmatrix}.$$

For a slightly complicated matrix

$$\mathbf{A} = \begin{pmatrix} t & -\omega \\ \omega & t \end{pmatrix},$$

we have

$$e^{\mathbf{A}} = \begin{pmatrix} e^t \cos\omega & -e^t \sin\omega \\ e^t \sin\omega & e^t \cos\omega \end{pmatrix}.$$

□

As you see, it is quite complicated but still straightforward to calculate the matrix exponentials. Fortunately, it can be easily done using a computer. By using the power expansions and the basic definition, we can prove the following useful identities

$$e^{t\mathbf{A}} \equiv \sum_{n=0}^{\infty} \frac{1}{n!}(t\mathbf{A})^n = \mathbf{I} + t\mathbf{A} + \frac{t^2}{2}\mathbf{A}^2 + ..., \tag{3.27}$$

$$\ln(\mathbf{IA}) \equiv \sum_{n=1}^{\infty} \frac{(-1)^{n-1}}{n!}\mathbf{A}^n = \mathbf{A} - \frac{1}{2}\mathbf{A}^2 + \frac{1}{3}\mathbf{A}^3 + ..., \tag{3.28}$$

$$e^{\mathbf{A}}e^{\mathbf{B}} = e^{\mathbf{A}+\mathbf{B}} \qquad (\text{if } \mathbf{AB} = \mathbf{BA}), \tag{3.29}$$

$$\frac{d}{dt}e^{t\mathbf{A}} = \mathbf{A}e^{t\mathbf{A}} = e^{t\mathbf{A}}\mathbf{A}, \tag{3.30}$$

$$(e^{\mathbf{A}})^{-1} = e^{-\mathbf{A}}, \tag{3.31}$$

$$\det(e^{\mathbf{A}}) = e^{\mathrm{tr}\mathbf{A}}. \tag{3.32}$$

## 3.5 Hermitian and Quadratic Forms

The matrices we have discussed so far are real matrices because all their elements are real. In general, the entries or elements of a matrix can be complex numbers, and the matrix becomes a complex matrix. For a matrix $\mathbf{A}$, its complex conjugate $\mathbf{A}^*$

is obtained by taking the complex conjugate of each of its elements. The Hermitian conjugate $\mathbf{A}^\dagger$ is obtained by taking the transpose of its complex conjugate matrix. That is to say, for

$$\mathbf{A} = \begin{pmatrix} a_{11}, & a_{12}, & \cdots \\ a_{21} & a_{21} & \cdots \\ \cdots & \cdots & \cdots \end{pmatrix}, \tag{3.33}$$

we have

$$\mathbf{A}^* = \begin{pmatrix} a_{11}^* & a_{12}^* & \cdots \\ a_{21}^* & a_{22}^* & \cdots \\ \cdots & \cdots & \cdots \end{pmatrix}, \tag{3.34}$$

and

$$\mathbf{A}^\dagger = (\mathbf{A}^*)^T = (\mathbf{A}^T)^* = \begin{pmatrix} a_{11}^* & a_{21}^* & \cdots \\ a_{12}^* & a_{22} & \cdots \\ \cdots & \cdots & \cdots \end{pmatrix}. \tag{3.35}$$

A square matrix $\mathbf{A}$ is called orthogonal if and only if $\mathbf{A}^{-1} = \mathbf{A}^T$. If a square matrix $\mathbf{A}$ satisfies $\mathbf{A}^* = \mathbf{A}$, it is said to be an Hermitian matrix. It is an anti-Hermitian matrix if $\mathbf{A}^* = -\mathbf{A}$. If the Hermitian matrix of a square matrix $\mathbf{A}$ is equal to the inverse of the matrix (or $\mathbf{A}^\dagger = \mathbf{A}^{-1}$), it is called a unitary matrix.

---

□ ***Example 3.3:*** *For a matrix*

$$\mathbf{B} = \begin{pmatrix} 2+i & 3-2i & 1 \\ e^{-i\pi} & 0 & 1-i\pi \end{pmatrix},$$

*its complex conjugate* $\mathbf{B}^*$ *and Hermitian conjugate* $\mathbf{B}^\dagger$ *are*

$$\mathbf{B}^* = \begin{pmatrix} 2-i & 3+2i & 1 \\ e^{i\pi} & 0 & 1+i\pi \end{pmatrix},$$

$$\mathbf{B}^\dagger = \begin{pmatrix} 2-i & e^{i\pi} \\ 3+2i & 0 \\ 1 & 1+i\pi \end{pmatrix} = (\mathbf{B}^*)^T.$$

*For the rotation matrix*

$$\mathbf{A} = \begin{pmatrix} \cos\theta & \sin\theta \\ -\sin\theta & \cos\theta \end{pmatrix},$$

*its inverse and transpose are*

$$\mathbf{A}^{-1} = \frac{1}{\cos^2\theta + \sin^2\theta} \begin{pmatrix} \cos\theta & -\sin\theta \\ \sin\theta & \cos\theta \end{pmatrix},$$

*and*

$$\mathbf{A}^T = \begin{pmatrix} \cos\theta & -\sin\theta \\ \sin\theta & \cos\theta \end{pmatrix}.$$

*Since* $\cos^2\theta + \sin^2\theta = 1$, *we have* $\mathbf{A}^T = \mathbf{A}^{-1}$. *Therefore, the original matrix* $\mathbf{A}$ *is orthogonal.* □

A very useful concept in engineering mathematics and computing is quadratic forms. For a real vector $\mathbf{q}^T = (q_1, q_2, q_3, ..., q_n$ and a real square matrix $\mathbf{A}$, a quadratic form $\psi(\mathbf{q})$ is a scalar function defined by

$$\psi(\mathbf{q}) = \mathbf{q}^T\mathbf{A}\mathbf{q}$$

$$= \begin{pmatrix} q_1 & q_2 & ... & q_n \end{pmatrix} \begin{pmatrix} A_{11} & A_{12} & ... & A_{1n} \\ A_{21} & A_{22} & ... & A_{2n} \\ ... & ... & ... & ... \\ A_{n1} & A_{n2} & ... & A_{nn} \end{pmatrix} \begin{pmatrix} q_1 \\ q_2 \\ \vdots \\ q_n \end{pmatrix}, \tag{3.36}$$

which can be written as

$$\psi(\mathbf{q}) = \sum_{i=1}^{n}\sum_{j=1}^{n} q_i A_{ij} q_j. \tag{3.37}$$

Since $\psi$ is a scalar, it should be independent of the coordinates. In the case of a square matrix $\mathbf{A}$, $\psi$ might be more easily evaluated in certain intrinsic coordinates $Q_1, Q_2, ...Q_n$. An important result concerning the quadratic form is that it can always be written through appropriate transformations as

$$\psi(\mathbf{q}) = \sum_{i=1}^{n} \lambda_i Q_i^2 = \lambda_1 Q_1^2 + \lambda_2 Q_2^2 + ...\lambda_n Q_n^2. \tag{3.38}$$

The natural extension of quadratic forms is the Hermitian form that is the quadratic form for complex Hermitian matrix $\mathbf{A}$.

Furthermore, the matrix **A** can be linear operators and functionals in addition to numbers.

---

□ ***Example 3.4:*** *For a vector* $\mathbf{q} = (q_1, q_2)$ *and the square matrix*

$$\mathbf{A} = \begin{pmatrix} 1 & -2 \\ -2 & 1 \end{pmatrix},$$

*we have a quadratic form*

$$\psi(\mathbf{q}) = \begin{pmatrix} q_1 & q_2 \end{pmatrix} \begin{pmatrix} 1 & -2 \\ -2 & 1 \end{pmatrix} \begin{pmatrix} q_1 \\ q_2 \end{pmatrix}$$

$$= q_1^2 - 4q_1q_2 + q_2^2.$$

---
□

## 3.6 Solution of linear systems

A linear system of $m$ equations for $n$ unknowns

$$a_{11}u_1 + a_{12}u_2 + ... + a_{1n}u_n = b_1,$$

$$a_{21}u_1 + a_{22}u_2 + ... + a_{2n}u_n = b_2,$$

$$\vdots \qquad \vdots$$

$$a_{m1}u_1 + a_{m2}u_2 + ... + a_{mn}u_n = b_n, \tag{3.39}$$

can be written in the compact form as

$$\begin{pmatrix} a_{11} & a_{12} & \dots & a_{1n} \\ a_{21} & a_{22} & \dots & a_{2n} \\ \vdots & \vdots & & \\ a_{m1} & a_{m2} & \dots & a_{mn} \end{pmatrix} \begin{pmatrix} u_1 \\ u_2 \\ \vdots \\ u_n \end{pmatrix} = \begin{pmatrix} b_1 \\ b_2 \\ \vdots \\ b_n \end{pmatrix}, \tag{3.40}$$

or simply

$$\mathbf{Au} = \mathbf{b}. \tag{3.41}$$

In the case of $m = n$, we multiply both sides by $\mathbf{A}^{-1}$ (this is only possible when $m = n$),

$$\mathbf{A^{-1}Au} = \mathbf{A^{-1}b}, \tag{3.42}$$

we obtain the solution

$$\mathbf{u} = \mathbf{A}^{-1}\mathbf{b}. \tag{3.43}$$

A special case of the above equation is when $\mathbf{b} = \lambda\mathbf{u}$, and this becomes an eigenvalue problem. An eigenvalue $\lambda$ and corresponding eigenvector $\mathbf{v}$ of a square matrix $\mathbf{A}$ satisfy

$$\mathbf{A}\mathbf{v} = \lambda\mathbf{v}, \tag{3.44}$$

or

$$(\mathbf{A} - \lambda\mathbf{I})\mathbf{v} = \mathbf{0}. \tag{3.45}$$

Any nontrivial solution requires

$$\begin{vmatrix} a_{11} - \lambda & a_{12} & \dots & a_{1n} \\ a_{21} & a_{22} - \lambda & \dots & a_{2n} \\ \vdots & \vdots & & \\ a_{n1} & a_{n2} & \dots & a_{nn} - \lambda \end{vmatrix} = 0, \tag{3.46}$$

which is equivalent to

$$\lambda^n + \alpha_{n-1}\lambda^{n-1} + ... + \alpha_0$$

$$= (\lambda - \lambda_1)(\lambda - \lambda_2)...(\lambda - \lambda_n) = 0. \tag{3.47}$$

In general, the characteristic equation has $n$ solutions. Eigenvalues have the interesting connections with the matrix,

$$\mathrm{tr}(\mathbf{A}) = \sum_{\mathrm{i}=1}^{\mathrm{n}} \mathrm{a_{ii}} = \lambda_1 + \lambda_2 + ... + \lambda_\mathrm{n}. \tag{3.48}$$

For a symmetric square matrix, the two eigenvectors for two distinct eigenvalues $\lambda_i$ and $\lambda_j$ are orthogonal $\mathbf{v}^{\mathbf{T}}\mathbf{v} = 0$.

Some useful identities involving eigenvalues and inverse of matrices are as follows:

$$(\mathbf{AB}...\mathbf{Z})^{-1} = \mathbf{Z}^{-1}...\mathbf{B}^{-1}\mathbf{A}^{-1}, \tag{3.49}$$

$$\mathbf{A}\mathbf{v}_i = \lambda_i\mathbf{v}_i, \qquad \lambda_i = \mathrm{eig}(\mathbf{A}). \tag{3.50}$$

$$\mathrm{eig}(\mathbf{AB}) = \mathrm{eig}(\mathbf{BA}), \tag{3.51}$$

$$\mathrm{tr}(\mathbf{A}) = \sum_i \mathbf{A}_{ii} = \sum_i \lambda_i. \tag{3.52}$$

$$\det(\mathbf{A}) = \Pi_i \lambda_i. \tag{3.53}$$

---

□ ***Example 3.5:*** *For a simple* $2 \times 2$ *matrix*

$$\mathbf{A} = \begin{pmatrix} 1 & 5 \\ 2 & 4 \end{pmatrix},$$

*its eigenvalues can be determined by*

$$\begin{vmatrix} 1-\lambda & 5 \\ 2 & 4-\lambda \end{vmatrix} = 0,$$

*or*

$$(1-\lambda)(4-\lambda) - 2 \times 5 = 0,$$

*which is equivalent to*

$$(\lambda+1)(\lambda-6) = 0.$$

*Thus, the eigenvalues are* $\lambda_1 = -1$ *and* $\lambda_2 = 6$. *The trace of* $\mathbf{A}$ *is* $tr(\mathbf{A}) = A_{11} + A_{22} = 1 + 4 = 5 = \lambda_1 + \lambda_2$.

*In order to obtain the eigenvector for each eigenvalue, we assume*

$$\mathbf{v} = \begin{pmatrix} v_1 \\ v_2 \end{pmatrix}.$$

*For the eigenvalue* $\lambda_1 = -1$, *we plug this into*

$$|\mathbf{A} - \lambda\mathbf{I}|\mathbf{v} = \mathbf{0},$$

*and we have*

$$\begin{vmatrix} 1-(-1) & 5 \\ 2 & 4-(-1) \end{vmatrix} \begin{pmatrix} v_1 \\ v_2 \end{pmatrix} = 0,$$

*or*

$$\begin{vmatrix} 2 & 5 \\ 2 & 5 \end{vmatrix} \begin{pmatrix} v_1 \\ v_2 \end{pmatrix} = 0,$$

*which is equivalent to*

$$2v_1 + 5v_2 = 0, \qquad \text{or} \qquad v_1 = -\frac{5}{2}v_2.$$

*This equation has infinite solutions, each corresponds to the vector parallel to the unit eigenvector. As the eigenvector should be normalized so that its modulus is unity, this additional condition requires*

$$v_1^2 + v_2^2 = 1,$$

*which means*

$$(\frac{-5v_2}{2})^2 + v_2^2 = 1.$$

*We have* $v_1 = -5/\sqrt{29}$, $v_2 = 2/\sqrt{29}$. *Thus, we have the first set of eigenvalue and eigenvector*

$$\lambda_1 = -1, \qquad \mathbf{v}_1 = \begin{pmatrix} -\frac{5}{\sqrt{29}} \\ \frac{2}{\sqrt{29}} \end{pmatrix}. \tag{3.54}$$

*Similarly, the second eigenvalue* $\lambda_2 = 6$ *gives*

$$\begin{vmatrix} 1-6 & 5 \\ 2 & 4-6 \end{vmatrix} \begin{pmatrix} v_1 \\ v_2 \end{pmatrix} = 0.$$

*Using the normalization condition* $v_1^2 + v_2^2 = 1$, *the above equation has the following solution*

$$\lambda_2 = 6, \qquad \mathbf{v}_2 = \begin{pmatrix} \frac{\sqrt{2}}{2} \\ \frac{\sqrt{2}}{2} \end{pmatrix}.$$

□

---

For a linear system $\mathbf{Au} = \mathbf{b}$, the solution $\mathbf{u} = \mathbf{A}^{-1}\mathbf{b}$ generally involves the inversion of a large matrix. The direct inversion becomes impractical if the matrix is very large (say, if $n > 1000$). Many efficient algorithms have been developed for solving such systems. Gauss elimination and LU decomposition are just two examples.

# Chapter 4

# Complex Variables

Although all the quantities are real variables in the physical world, however, it is sometimes easy or even necessary to use complex variables in mathematics and engineering. In fact, the techniques based on complex variables are among the most powerful methods for mathematical analysis and solutions of mathematical models.

## 4.1 Complex Numbers and Functions

Mathematically speaking, a complex number $z$ is a generalized set or the order pair of two real numbers $(a, b)$, written in the form of

$$z = a + ib, \qquad i^2 = -1, \qquad a, b \in \mathcal{R}, \tag{4.1}$$

which consists of the real part $\Re(z) = a$ and the imagery part $\Im(z) = b$. It can also be written as the order pair of real numbers using the notation $(a, b)$. The addition and substraction of two complex numbers are defined as

$$(a + ib) \pm (c + id) = (a \pm c) + i(b \pm d). \tag{4.2}$$

The multiplication and division of two complex numbers are in the similar way as expanding polynomials

$$(a + ib) \cdot (c + id) = (ac - bd) + i(ad + bc). \tag{4.3}$$

and

$$\frac{a+ib}{c+id} = \frac{ac+bd}{c^2+d^2} + i\frac{bc-ad}{c^2+d^2}. \tag{4.4}$$

Two complex numbers are equal $a+ib = c+id$ if and only if $a = c$ and $b = d$. The complex conjugate or simply conjugate $\bar{z}$ (also $z^*$) of $z = a + ib$ is defined as

$$\bar{z} = a - ib. \tag{4.5}$$

The order pair $(a, b)$, similar to a vector, implies that a geometrical representation of a complex number $a + ib$ by the point in an ordinary Euclidean plane with $x$-axis being the real axis and $y$-axis being the imaginary axis ($iy$). This plane is called the complex plane. This representation is often called the Argand diagram (see Figure 4.1). The vector representation starts from $(0, 0)$ to the point $(a, b)$. The length of the vector is called the magnitude or modulus or the absolute value of the complex number

$$r = |z| = \sqrt{a^2 + b^2}. \tag{4.6}$$

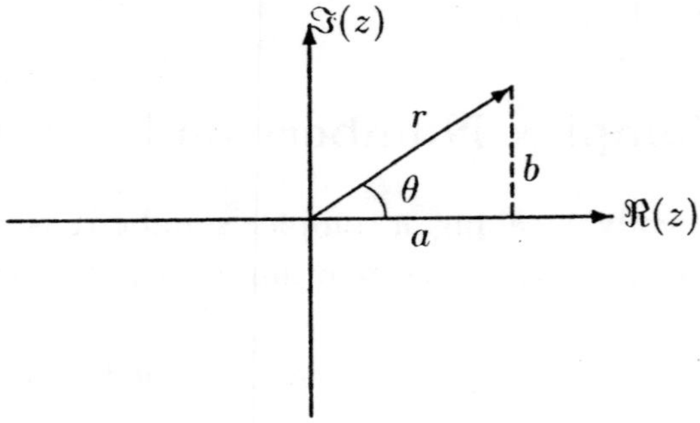


Figure 4.1: Polar representation of a complex number.

The angle $\theta$ that the vector makes with the positive real axis is called the argument (see Fig 4.1),

$$\theta = \arg z. \tag{4.7}$$

In fact, we may replace $\theta$ by $\theta + 2n\pi$ ($n \in \mathcal{N}$). The value range $-\pi < \theta \le \pi$ is called the principal argument of $z$, and it is usually denoted as Arg$z$. In the complex plane, the complex number can be written as

$$z = re^{i\theta} = r\cos(\theta) + ir\sin(\theta). \tag{4.8}$$

This polar form of $z$ and its geometrical representation can result in the Euler's formula which is very useful in the complex analysis

$$e^{i\theta} = \cos(\theta) + i\sin(\theta). \tag{4.9}$$

The Euler formula can be proved using the power series. For any $z \in \mathcal{C}$, we have the power series

$$e^z = 1 + z + \frac{z^2}{2!} + ... + \frac{z^n}{n!} + ..., \tag{4.10}$$

and for a special case $z = i\theta$, we have

$$e^{i\theta} = 1 + i\theta - \frac{\theta^2}{2!} + \frac{i\theta^3}{3!} - ...,$$

$$= (1 - \frac{\theta^2}{2!} + ...) + i(\theta - \frac{\theta^3}{3!} + ...). \tag{4.11}$$

Using the power series

$$\sin\theta = \theta - \frac{\theta^3}{3!} + \frac{\theta^5}{5!} - ..., \tag{4.12}$$

and

$$\cos\theta = 1 - \frac{\theta^2}{2!} + \frac{\theta^4}{4!} - ..., \tag{4.13}$$

we get the well-know Euler's formula or Euler's equation

$$e^{i\theta} = \cos\theta + i\sin\theta. \tag{4.14}$$

For $\theta = \pi$, this leads to a very interesting formula

$$e^{i\pi} + 1 = 0. \tag{4.15}$$

If we replace $\theta$ by $-\theta$, the Euler's formula becomes

$$e^{-i\theta} = \cos(-\theta) + i\sin(-\theta) = \cos\theta - i\sin\theta. \tag{4.16}$$

Adding this equation to (4.14), we have

$$e^{i\theta} + e^{-i\theta} = 2\cos\theta, \tag{4.17}$$

or

$$\cos\theta = \frac{e^{i\theta} + e^{-i\theta}}{2}. \tag{4.18}$$

Similarly, by deducting (4.16) from (4.14), we get

$$\sin\theta = \frac{e^{i\theta} - e^{-i\theta}}{2i}. \tag{4.19}$$

For two complex numbers $z_1 = r_1 e^{i\alpha_1}$ and $z_2 = r_2 e^{i\alpha_2}$, it is straightforward to show that

$$z_1 z_2 = r_1 r_2 e^{i(\alpha_1+\alpha_2)} = r_1 r_2[\cos(\alpha_1 + \alpha_2) + i\sin(\alpha_1 + a_2)], \tag{4.20}$$

which can easily be extended to get the well-known de Moivre's formula

$$[\cos(\theta) + i\sin(\theta)]^n = \cos(n\theta) + i\sin(n\theta). \tag{4.21}$$

---

□ ***Example 4.1:*** *Find $z^4$ if $z = 1 + \sqrt{3}i$. We can evaluate it by direct calculation*

$$z^4 = (1 + \sqrt{3}i)^4 = [(1 + \sqrt{3}i)^2]^2 = [1 - 3 + 2\sqrt{3}i]^2$$

$$= 2^2(-1 + \sqrt{3}i)^2 = 4(1 - 3 - 2\sqrt{3}) = -8 - 8\sqrt{3}i.$$

*We can also use Moivre's formula. The modulus of $z$ is $r = |z| = \sqrt{1^2 + \sqrt{3}^2} = 2$. The argument $\theta = \tan^{-1}\frac{\sqrt{3}}{1} = \pi/3 = 60°$. Thus, $z = 2e^{\pi/3}$. We now have*

$$z^4 = 2^4 e^{4\pi/3} = 16(\cos\frac{4\pi}{3} + i\sin\frac{4\pi}{3})$$

$$= 16(-\frac{1}{2} - \frac{\sqrt{3}}{2}i) = -8 - 8\sqrt{3}i,$$

*which is exactly the same result as we obtained earlier. The second method becomes much quicker if you want to evaluate (say) $z^{100}$.*

□

---

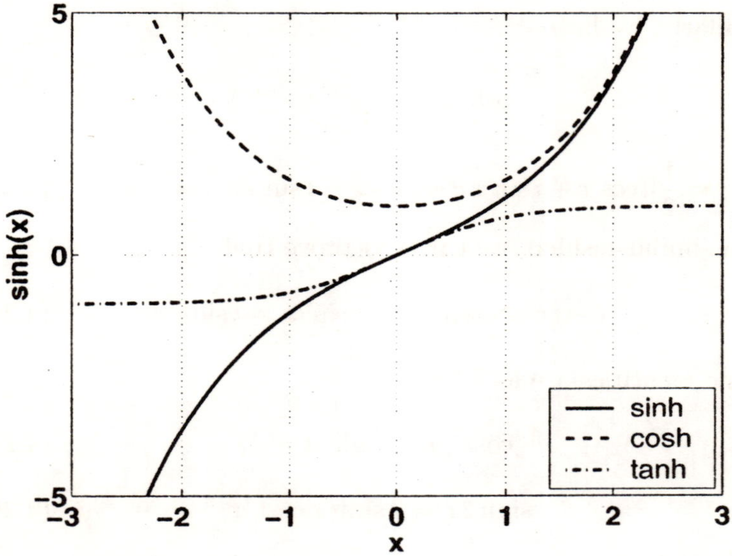


Figure 4.2: Hyperbolic functions.

## 4.2 Hyperbolic Functions

Hyperbolic functions occur in many applications and they can be thought as the complex analogues of trigonometric functions. The fundamental definitions are

$$\sinh x = \frac{e^x - e^{-x}}{2}, \qquad \cosh x = \frac{e^x + e^{-x}}{2}, \tag{4.22}$$

and

$$\tanh x = \frac{\sinh x}{\cosh x}, \qquad \coth x = \frac{1}{\tanh x}. \tag{4.23}$$

Figure 4.2 shows the variation of sinh, cosh, and tanh. If we replace $x$ by $ix$ and use Euler's formula, then we have

$$\sinh ix = \frac{e^{ix} - e^{-ix}}{2}$$

$$= \frac{1}{2}[(\cos x + i\sin x) - (\cos x - i\sin x)] = i\sin x. \tag{4.24}$$

Similarly, we have

$$\cosh ix = \frac{1}{2}(e^{ix} + e^{-ix})$$

$$= \frac{1}{2}[(\cos x + i \sin x) + (\cos x - i \sin x)] = \cos x. \tag{4.25}$$

In a similar fashion, we can also prove that

$$\cos ix = \cosh x, \qquad \sin ix = i \sinh x. \tag{4.26}$$

Some identities are as follows:

$$\cosh^2 x - \sinh^2 x = 1, \tag{4.27}$$

$$\sinh 2x = 2 \sinh x \cosh x, \tag{4.28}$$

and

$$\cosh 2x = \sinh^2 x + \cosh^2 x. \tag{4.29}$$

---

□ ***Example 4.2:*** *Prove that $\cosh^2 x - \sinh^2 x = 1$. From the definitions, we have*

$$\cosh^2 x = \frac{1}{4}(e^x + e^{-x})^2 = \frac{1}{4}(e^{2x} + 2 + e^{-2x}),$$

*and*

$$\sinh^2 x = \frac{1}{4}(e^x - e^{-x})^2 = \frac{1}{4}(e^{2x} - 2 + e^{-2x}).$$

*Thus, we have*

$$\cosh^2 x - \sinh^2 x = \frac{1}{4}[(e^{2x} + 2 + e^{-2x}) - (e^{2x} - 2 + e^{-2x})]$$

$$= \frac{1}{4}[2 - (-2)] = 1.$$

□

---

The inverses of hyperbolic functions are defined in a similar way as trigonometric functions. For example, $y = \cosh x$, its inverse is defined as $x = \cosh^{-1} y$. From the basic definitions, we have

$$\sinh x + \cosh x = e^x. \tag{4.30}$$

Using $\sinh x = \sqrt{\cosh^2 x - 1}$, we have

$$\sqrt{\cosh^2 x - 1} + \cosh x = e^x, \tag{4.31}$$

or

$$\sqrt{y^2 - 1} + y = e^x, \tag{4.32}$$

which gives

$$x = \cosh^{-1} y = \ln(\sqrt{y^2 - 1} + y). \tag{4.33}$$

## 4.3 Analytic Functions

### Analytic Functions

Any function of real variables can be extended to the function of complex variables in the same form while treating the real numbers $x$ as $x + i0$. For example, $f(x) = x^2, x \in \mathcal{R}$ becomes $f(z) = z^2, z \in \mathcal{C}$. Any complex function $f(z)$ can be written as

$$f(z) = f(x + iy) = \Re(f(z)) + i\Im(f(z))$$
$$= u(x, y) + iv(x, t), \tag{4.34}$$

where $u(x, y)$ and $v(x, y)$ are real-valued functions of two real variables.

A function $f(z)$ is called analytic at $z_0$ if $f'(z)$ exists for all $z$ in some $\epsilon$−neighborhood of $z_0$, that is to say, it is differentiable in some open disk $|z - z_0| < \epsilon$. If $f(z) = u + iv$ is analytic at every point in a domain $\Omega$, then $u(x, y)$ and $v(x, y)$ satisfying the Cauchy-Riemann equations

$$\frac{\partial u}{\partial x} = \frac{\partial v}{\partial y}, \qquad \frac{\partial u}{\partial y} = -\frac{\partial v}{\partial x}. \tag{4.35}$$

Conversely, if $u$ and $v$ of $f(z) = u + iv$ satisfy the Cauchy-Riemann equation at all points in a domain, then the complex function $f(z)$ is analytic in the same domain. For example, the elementary power function $w = z^n, (n > 1)$ is analytic on the whole plane, $w = \rho e^{i\phi}$, $z = re^{i\theta}$, then

$$\rho = r^n, \phi = n\theta. \tag{4.36}$$

The logarithm is also an elementary function $w = \ln z$

$$\ln z = \ln |z| + i \arg(z) = \ln r + i(\theta + w\pi k), \tag{4.37}$$

which has infinitely many values, due to the multiple values of $\theta$, with the difference of $2\pi ik (k = 0, \pm 1, \pm 2, ...)$. If we use the principal argument Arg$z$, then we have the principal logarithm function

$$\mathrm{Ln}(z) = \ln |z| + \mathrm{Arg} z. \tag{4.38}$$

If we differentiate the Cauchy-Riemann equations, we have $\partial^2 u/\partial x \partial y = \partial^2 u/\partial y \partial x$. After some calculations, we can reach the following theorem. For given analytic function $f(z) = u + iv$, then both $u$ and $v$ satisfy the Laplace equations

$$\frac{\partial^2 u}{\partial x^2} + \frac{\partial^2 v}{\partial y^2} = 0, \qquad \frac{\partial^2 v}{\partial x^2} + \frac{\partial^2 v}{\partial y^2} = 0. \tag{4.39}$$

This is to say, both real and imaginary parts of an analytic function are harmonic.

A very interesting analytical function is the Riemann zeta-function $\zeta(s)$, which is defined by

$$\zeta(s) = \sum_{n=1}^{\infty} \frac{1}{n^s}, \tag{4.40}$$

where $s$ is a complex number with its real part more than unity. That is $s \in C$ and $\Re(s) > 1$. This function (infinite series) is analytic, and it can be extended for all complex numbers $s \neq 1$. For example,

$$\zeta(2) = 1 + \frac{1}{2^2} + \frac{1}{3^2} + ... = \frac{\pi^2}{6}, \tag{4.41}$$

but

$$\zeta(1) = 1 + \frac{1}{2} + \frac{1}{3} + \frac{1}{4} + ... = \infty. \tag{4.42}$$

This $\zeta(s)$ function has trivial zeros $s = -2, -4, -6, ...$ and it also has non-trivial zeros.

There is a famous unsolved problem, called the Riemann hypothesis, related to this function. The Riemann hypothesis conjectured by Bernhard Reimann in 1859 states that all real parts of any non-trial zero of the Riemann zeta-function $\zeta(s)$ are $\frac{1}{2}$. That is to say, all the non-trivial zeros should lie on a straight line $s = \frac{1}{2} + iy$. This is a-million-dollar open problem as the Clay Mathematical Institute in 2000 offered a million dollars to search for a proof, and yet it still remains unsolved.

### Laurent Series

For an analytic function $p(z)$, one of important properties is the singularity such as the pole. If $p(z)$ can be written as

$$p(z) = \frac{q(z)}{(z - z_0)^n}, \tag{4.43}$$

where $n > 0$ is a positive integer while $q(z) \neq 0$ is analytic everywhere in the neighbourhood containing $z = z_0$, we say that $p(z)$ has a pole of order $n$ at $z = z_0$. The above definition is equivalent to say that the following limit is finite

$$\lim_{z \to z_0} [p(z)(z - z_0)^n] = \zeta. \qquad ||\zeta|| < \infty, \ \zeta \in \mathcal{C}. \tag{4.44}$$

Any analytic function $f(z)$ can be expanded in terms of the Taylor series

$$f(z) = \sum_{k=0}^{\infty} \frac{f^{(k)}}{k!}(z - z_0)^k = \sum_{k=0}^{\infty} \alpha_k (z - z_0)^k. \tag{4.45}$$

This expansion is valid inside the analytic region. However, if the function $f(z)$ has a pole of order $n$ at $z = z_0$ and it is analytic everywhere except at the pole, we can then expand the function $p(z) = (z - z_0)^n f(z)$ in the standard Taylor expansion.

This means that original function $f(z)$ can be written as a power series

$$f(z) = \frac{\alpha_{-n}}{(z-z_0)^n} + ... + \frac{\alpha_{-1}}{(z-z_0)}$$

$$\alpha_0(z-z_0) + ... + \alpha_k(z-z_0)^k + ..., \tag{4.46}$$

which is called a Laurent series, and it is an extension of the Taylor series. In this series, it is often assumed that $\alpha_{-n} \neq 0$. The terms with the inverse powers $a_{-n}/(z-z_0)^n + ... + a_{-1}/(z-z_0)$ are called the principal part of the series, while the usual terms $a_0(z-z_0) + ... + \alpha_k(z-z_0)^k + ...$ are called the analytic part.

Furthermore, the most important coefficient is probably $\alpha_{-1}$ which is called the residue of $f(z)$ at the pole $z = z_0$. In general, the Laurent series can be written as

$$f(z) = \sum_{k=-n}^{\infty} \alpha_k(z-z_0)^k, \tag{4.47}$$

where $n$ may be extended to include an infinite number of terms $n \to -\infty$.

## 4.4 Complex Integrals

Given a function $f(z)$ that is continuous on a piecewise smooth curve $\Gamma$, then the integral over $\Gamma$, $\int_\Gamma f(z)dz$, is called a contour or line integral of $f(z)$. This integral has similar properties as the real integral

$$\int_\Gamma [\alpha f(z) + \beta g(z)]dz = \alpha \int_\Gamma f(z)dz + \beta \int_\Gamma g(z)dz. \tag{4.48}$$

If $F(z)$ is analytic and $F'(z) = f(z)$ is continuous along a curve $\Gamma$, then

$$\int_a^b f(z)dz = F[z(b)] - F[z(a)]. \tag{4.49}$$

## Cauchy's Integral Theorem

We say a path is simply closed if its end points and initial points coincide and the curve does not cross itself. For an analytic function $f(z) = u(x, y) + iv(x, y)$, the integral on a simply closed path

$$I = \int_\Gamma f(z)dz = \int_\Gamma (u + iv)(dx + idy)]$$

$$= \int_\Gamma (udx - vdy) + i \int_\Gamma (vdx + udy). \quad (4.50)$$

By using the Green theorem, this becomes

$$I = \int_\Omega (-\frac{\partial u}{\partial y} - \frac{\partial v}{\partial x})dxdy + i \int_\Omega (\frac{\partial u}{\partial x} - \frac{\partial v}{\partial y})dxdy. \quad (4.51)$$

From the Cauchy-Riemann equations, we know that both integrals are zero. Thus, we have Cauchy's Integral Theorem, which states that the integral of any analytic function $f(z)$ on a simply closed path $\Gamma$ in a simply connected domain $\Omega$ is zero. That is

$$\int_\Gamma f(z)dz = 0.$$

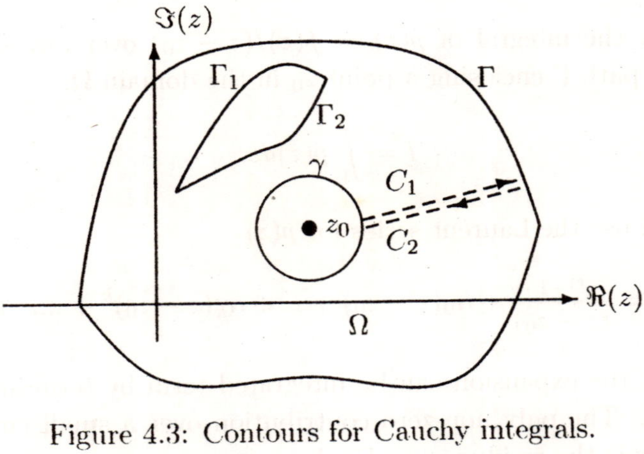


Figure 4.3: Contours for Cauchy integrals.

This theorem is very important as it has interesting consequences. If the close path is decomposed into two paths with reverse directions $\Gamma_1$ and $\Gamma_2$ (see Figure 4.3), then $\Gamma_1$ and $-\Gamma_2$ form a close path, which leads to

$$\int_{\Gamma_1} f(z)dz = \int_{\Gamma_2} f(z)dz. \tag{4.52}$$

That is to say that the integrals over any curve between two points are independent of path. This property becomes very useful for evaluation of integrals. In fact, this can be extended to the integrals over two closed paths $\Gamma$ and $\gamma$ such that $\gamma$ is a very small circular path inside $\Gamma$. Using a small cut with two curves $C_1$ and $C_2$ so that these two curves combine with $\Gamma$ and $\gamma$ form a closed contour (see Figure 4.3), the Cauchy integral theorem implies that

$$\int_{\Gamma} f(z)dz = \int_{\gamma} f(z)dz, \tag{4.53}$$

since the contribution from the cut is zero.

For an analytic function with a pole, we can make the contour $\gamma$ sufficiently small to enclose just around the pole, and this makes the calculation of the integral much easier in some cases.

For the integral of $p(z) = f(z)/(z - z_0)$ over any simply closed path $\Gamma$ enclosing a point $z_0$ in the domain $\Omega$,

$$I = \int_{\Gamma} p(z)dz, \tag{4.54}$$

we can use the Laurent series for $p(z)$

$$p(z) = \frac{\alpha_{-1}}{(z - z_0)} + \alpha_0(z - z_0) + ... + \alpha_k(z - z_0)^k + ..., \tag{4.55}$$

so that the expansion can be integrated term by term around a path. The only non-zero contribution over a small circular contour is the residue $\alpha_{-1}$. We have

$$I = \int_{\Gamma} p(z)dz = 2\pi i\alpha_{-1} = 2\pi i \, \mathrm{Res}[p(z)]\Big|_{z_0}, \tag{4.56}$$

which can be written in terms of $f(z)$ as

$$\frac{1}{2\pi i}\oint_\Gamma \frac{f(z)}{z-z_0}dz = f(z_0). \tag{4.57}$$

Similarly, this can be extended for higher derivatives, and we have

$$\oint_\Gamma \frac{f(z)}{(z-z_0)^{n+1}}dz = \frac{2\pi i f^{(n)}(z_0)}{n!}.$$

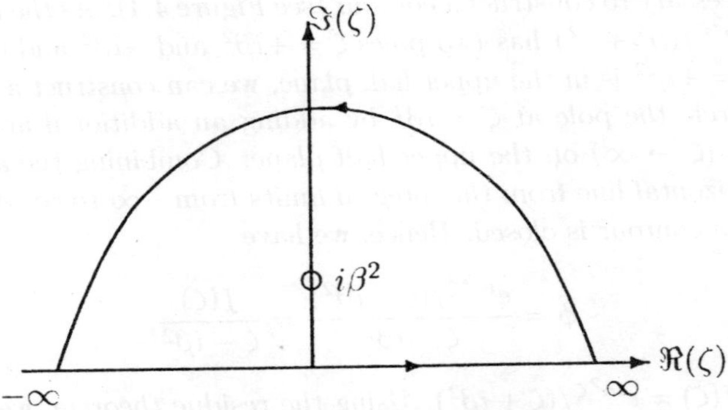

Figure 4.4: Contour for the integral $I(\alpha, \beta)$.

## Residue Theorem

For any analytic $f(z)$ function in a domain $\Omega$ except isolated singularities at finite points $z_1, z_2, ..., z_N$, the residue theorem states

$$\oint_\Gamma f(z)dz = 2\pi i \sum_{k=1}^{N} \text{Res} f(z)|_{z_k},$$

where $\Gamma$ is a simple closed path enclosing all these isolated points. If $f(z)$ has a pole of order $N$ at $z_0$, the following formula gives a quick way to calculate the residue

$$\text{Res} f(z)|_{z_0} = \frac{1}{(N-1)!}\lim_{z\to z_0}\frac{d^{N-1}[(z-z_0)^N f(z)]}{dz^{N-1}}. \tag{4.58}$$

The residue theorem serves a powerful tool for calculating some real integrals and summation of series, especially when the integrand is a function of sin and cos that can be changed into a complex integral. The real integral $\int_{-\infty}^{\infty} \psi(x)dx$ becomes $2\pi i$ multiplying the sum of the residues of $\psi(x)$ at the poles in the upper half-space.

---

□ ***Example 4.3:*** *In order to evaluate the integral*

$$I(\alpha, \beta) = \int_{\infty}^{\infty} \frac{e^{i\alpha^2\zeta}}{x^2 + \beta^4} d\zeta,$$

*it is necessary to construct a contour (see Figure 4.4). As the function $\phi = e^{i\alpha^2\zeta}/(\beta^4 + \zeta^2)$ has two poles $\zeta = +i\beta^2$ and $-i\beta^2$ and only one pole $\zeta = +i\beta^2$ is in the upper half plane, we can construct a contour to encircle the pole at $\zeta = i\beta^2$ by adding an additional arc at the infinity ($\zeta \to \infty$) on the upper half plane. Combining the arc with the horizontal line from the integral limits from $-\infty$ to $\infty$ along the $\zeta$-axis, a contour is closed. Hence, we have*

$$\phi = \frac{e^{i\alpha^2\zeta}/(\zeta + i\beta^2)}{\zeta - i\beta^2} = \frac{f(\zeta)}{\zeta - i\beta^2},$$

*where $f(\zeta) = e^{i\alpha^2\zeta}/(\zeta + i\beta^2)$. Using the residue theorem, we have*

$$I = 2\pi i[f(\zeta = i\beta^2)] = 2\pi i \frac{e^{-\alpha^2\beta^2}}{i\beta^2 + i\beta^2} = \pi \frac{e^{-\alpha^2\beta^2}}{\beta^2}.$$

*In a special case when $\alpha = 0$, we have*

$$\int_{-\infty}^{\infty} \frac{1}{\zeta^2 + \beta^4} d\zeta = \frac{\pi}{\beta^2}.$$

□

---

Another important topic in complex variables is the conformal mapping. The essence of a conformal mapping

$$w = f(z), \qquad z, w \in C, \tag{4.59}$$

is that this mapping preserves the angles between curves and their orientations. One of the widely used mappings is Möbius linear fractional transformation

$$w = \frac{\alpha z + \beta}{\gamma z + \delta}, \qquad \begin{vmatrix} \alpha & \beta \\ \gamma & \delta \end{vmatrix} \neq 0. \tag{4.60}$$

By choosing the appropriate coefficients $\alpha, \beta, \gamma, \delta \in \mathcal{C}$, this mapping can include all major geometrical transformations such as translations, rotations, inversion, and expansions and contractions. Conformal mappings are useful in solving steady-state problems involving harmonic functions by transforming the problem from a complicated geometrical domain to a regular domain such as circles and rectangles, and subsequently the techniques based on conformal mapping are widely used in solving Laplace's equation in engineering.

# Chapter 5

# Ordinary Differential Equations

Most mathematical models in engineering are formulated in terms of differential equations. If the variables or quantities (such as velocity, temperature, pressure) change with other independent variables such as spatial coordinates and time, their relationship can in general be written as a differential equation or even a set of differential equations.

## 5.1 Introduction

An ordinary differential equation (ODE) is a relationship between a function $y(x)$ of an independent variable $x$ and its derivatives $y'$, $y''$, ..., $y^{(n)}$. It can be written in a generic form

$$\Psi(x, y, y', y'', ..., y^{(n)}) = 0. \tag{5.1}$$

The solution of the equation is a function $y = f(x)$, satisfying the equation for all $x$ in a given domain $\Omega$.

The order of the differential equation is equal to the order $n$ of the highest derivative in the equation. Thus, the Riccati equation:

$$y' + a(x)y^2 + b(x)y = c(x), \tag{5.2}$$

is a first order ODE, and the following equation of Euler-type

$$x^2y'' + a_1xy' + a_0y = 0, \tag{5.3}$$

is a second order. The degree of the equation is defined as the power to which the highest derivative occurs. Therefore, both Riccati equation and Euler equation are of the first degree. An equation is called linear if it can be arranged into the form

$$a_n(x)y^{(n)} + ... + a_1(x)y' + a_0(x)y = \phi(x), \tag{5.4}$$

where all the coefficients depend on $x$ only, not on $y$ or any derivatives. If any of the coefficients is a function of $y$ or any of its derivatives, then the equation is nonlinear. If the right hand side is zero or $\phi(x) \equiv 0$, the equation is homogeneous. It is called nonhomogeneous if $\phi(x) \neq 0$.

The solution of an ordinary differential equation is not always straightforward, and it is usually very complicated for nonlinear equations. Even for linear equations, the solutions can only be obtained for a few simple types. The solution of a differential equation generally falls into three types: closed form, series form and integral form. A closed form solution is the type of solution that can be expressed in terms of elementary functions and some arbitrary constants. Series solutions are the ones that can be expressed in terms of a series when a closed-form is not possible for certain type of equations. The integral form of solutions or quadrature is sometimes the only form of solutions that are possible. If all these forms are not possible, the alternatives are to use approximate and numerical solutions.

## 5.2 First Order ODEs

### 5.2.1 Linear ODEs

The general form of a first order linear differential equation can be written as

$$y' + a(x)y = b(x). \tag{5.5}$$

This equation is always solvable using the integrating factor and it has a closed form solution.

Multiplying both sides of the equation by $\exp[\int a(x)dx]$, which is often called the integrating factor, we have

$$y'e^{\int a(x)dx} + a(x)ye^{\int a(x)dx} = b(x)e^{\int a(x)dx}, \tag{5.6}$$

which can be written as

$$[ye^{\int a(x)dx}]' = b(x)e^{\int a(x)dx}. \tag{5.7}$$

By simple integration, we have

$$ye^{\int a(x)dx} = \int b(x)e^{\int a(x)dx}dx + C. \tag{5.8}$$

So its solution becomes

$$y(x) = e^{-\int a(x)dx}\int b(x)e^{\int a(x)dx}dx + Ce^{-\int a(x)dx}, \tag{5.9}$$

where $C$ is an integration constant. The integration constant can be determined if extra requirements are given, and these extra requirements are usually the initial condition when time is zero or boundary conditions at some given points which are at the domain boundary. However, the classification of conditions may also depend on the meaning of the independent $x$. If $x$ is spatial coordinate, then $y(x = 0) = y_0$ is boundary condition at $x = 0$. However, if $x = t$ means time, then $y(t = 0) = y_0$ can be thought of as the initial condition at $t = 0$. Nevertheless, one integration constant usually requires one condition to determine it.

---

□ ***Example 5.1:*** *We now try to solve the ordinary differential equation* $\frac{dy}{dt} + ty(t) = -t$ *with an initial condition* $y(0) = 0$. *As* $a(t) = t, b(t) = -t$, *its general solution is*

$$y(t) = e^{-\int tdt}\int(-t)e^{\int tdt}dt + Ce^{-\int tdt}$$

$$= -e^{-\frac{t^2}{2}}\int te^{\frac{t^2}{2}}dt + Ce^{-\frac{t^2}{2}}$$

$$= -e^{\frac{t^2}{2}} e^{\frac{t^2}{2}} + Ce^{-\frac{t^2}{2}} = -1 + Ce^{-\frac{t^2}{2}}.$$

*From the initial condition* $y(0) = 0$ *at* $t = 0$*, we have*

$$0 = -1 + C, \qquad \text{or} \qquad C = 1.$$

*Thus, the solution becomes*

$$y(t) = e^{-\frac{t^2}{2}} - 1.$$

□

---

### 5.2.2 Nonlinear ODEs

For some nonlinear first order ordinary differential equations, sometimes a transform or change of variables can convert it into the standard first order linear equation (5.5). For example, the Bernoulli's equation can be written in the generic form

$$y' + p(x)y = q(x)y^n, \qquad n \neq 1. \tag{5.10}$$

In the case of $n = 1$, it reduces to a standard first order linear ordinary differential equation. By dividing both sides by $y^n$ and using the change of variables

$$u(x) = \frac{1}{y^{n-1}}, \qquad u' = \frac{(1-n)y'}{y^n}, \tag{5.11}$$

we have

$$u' + (1-n)p(x)u = (1-n)q(x), \tag{5.12}$$

which is a standard first order linear differential equation whose general solution is given earlier in this section.

---

□ ***Example 5.2:*** *To solve* $y'(x) + xy = y^{20}$*, we first use* $u(x) = 1/y^{19}$*, and we* $u' = -19y'/y^{20}$*. The original equation becomes*

$$u' - 19xu = -19,$$

*whose general solution is*

$$u(x) = Ae^{19x} + 1.$$

Therefore, the solution to the original equation becomes

$$y^{19} = \frac{1}{Ae^{19x}+1}, \qquad \text{or} \qquad y = (Ae^{19x}+1)^{-\frac{1}{19}}.$$

□

## 5.3 Higher Order ODEs

Higher order ODEs are more complicated to solve even for the linear equations. For the special case of higher-order ODEs where all the coefficients $a_n, ..., a_1, a_0$ are constants,

$$a_n y^{(n)} + ... + a_1 y' + a_0 y = f(x), \tag{5.13}$$

its general solution $y(x)$ consists of two parts: the complementary function $y_c(x)$ and the particular integral or particular solution $y_p^*(x)$. We have

$$y(x) = y_c(x) + y_p^*(x). \tag{5.14}$$

### 5.3.1 General Solution

The complementary function is the solution of the linear homogeneous equation with constant coefficients and can be written in a generic form

$$a_n y_c^{(n)} + a_{n-1} y_c^{(n-1)} + ... + a_1 y_c' + a_0 = 0. \tag{5.15}$$

Assuming $y = Ae^{\lambda x}$, we get the polynomial equation of characteristics

$$a_n \lambda^n + a_{n-1}\lambda^{(n-1)} + ... + a_1 \lambda + a_0 = 0, \tag{5.16}$$

which has $n$ roots in general. Then, the solution can be expressed as the summation of various terms $y_c(x) = \sum_{k=1}^{n} c_k e^{\lambda_k x}$ if the polynomial has $n$ distinct zeros $\lambda_1, ...\lambda_n$. For complex roots, and complex roots always occur in pairs $\lambda = r \pm i\omega$, the corresponding linearly independent terms can then be replaced by $e^{rx}[A\cos(\omega x) + B\sin(\omega x)]$.

The particular solution $y_p^*(x)$ is any $y(x)$ that satisfies the original inhomogeneous equation (5.13). Depending on the form of the function $f(x)$, the particular solutions can take various forms. For most of the combinations of basic functions such as $\sin x, \cos x$, $e^{kx}$, and $x^n$, the method of the undetermined coefficients is widely used. For $f(x) = \sin(\alpha x)$ or $\cos(\alpha x)$, then we can try $y_p^* = A\sin\alpha x + B\sin\alpha x$. We then substitute it into the original equation (5.13) so that the coefficients $A$ and $B$ can be determined. For a polynomial $f(x) = x^n (n = 0, 1, 2, ...., N)$, we then try $y_p^* = A + Bx + Cx^2 + ... + Qx^n$ (polynomial). For $f(x) = e^{kx}x^n$, $y_p^* = (A + Bx + Cx^2 + ...Qx^n)e^{kx}$. Similarly, $f(x) = e^{kx}\sin\alpha x$ or $f(x) = e^{kx}\cos\alpha x$, we can use $y_p^* = e^{kx}(A\sin\alpha x + B\cos\alpha x)$. More general cases and their particular solutions can be found in various textbooks.

---

□ ***Example 5.3:*** *In order to solve the equation $y'''(x) - 2y''(x) - y'(x) + 2y(x) = \sin x$, we have to find its complementary function $y_c(x)$ and its particular integral $y^*(x)$. we first try to solve its complementary equation or homogeneous equation*

$$y'''(x) - 2y''(x) - y'(x) + 2y(x) = 0.$$

*Assuming that $y = Ae^{\lambda x}$, we have the characteristic equation*

$$\lambda^3 - 2\lambda^2 - \lambda + 2 = 0,$$

*or*

$$(\lambda - 1)(\lambda + 1)(\lambda - 2) = 0.$$

*Thus, three basic solutions are $e^x, e^{-x}$ and $e^{2x}$. The general complementary function becomes*

$$y_c = Ae^x + Be^{-x} + Ce^{2x}.$$

*As the function $f(x) = \sin x$, thus we can assume that the particular integral takes the form $y^*(x) = a\sin x + b\cos x$. Substituting this into the original equation, we have*

$$(-a\cos x + b\sin x) - 2(-a\sin x - b\cos x)$$

$$-(a\cos x - b\sin x) + 2(a\sin x + b\cos x) = \sin x,$$

*or*

$$(b + 2a + b + 2a - 1)\sin x + (-a + 2b - a + 2b)\cos x = 0.$$

*Thus, we have*

$$4a + 2b = 1, -2a + 4b = 0,$$

*whose solution becomes*

$$a = \frac{1}{5}, \qquad b = \frac{1}{10}.$$

*Now the particular integral becomes*

$$y^*(x) = \frac{1}{5}\sin x + \frac{1}{10}\cos x.$$

*Finally, the general solution*

$$y = \frac{1}{5}\sin x + \frac{1}{10}\cos x + Ae^{x} + Be^{-x} + Ce^{2x}.$$

□

---

The methods for finding particular integrals work for most cases. However, there are some problems in the case when the right-hand side of the differential equation has the same form as part of the complementary function. In this case, the trial function should include one higher order term obtained by multiplying the standard trial function by the lowest integer power of $x$ so that the product does not appear in any term of the complementary function. Let us see an example.

---

□ ***Example 5.4:*** *Consider the equation*

$$y''(x) - 3y'(x) + 2y(x) = e^{x}.$$

*Using $y(x) = Ke^{\lambda x}$, we have the characteristic equation*

$$\lambda^2 - 3\lambda + 2 = 0.$$

*Its complementary function is*

$$y_c = Ae^{x} + Be^{2x}$$

*As the right hand side $f(x) = e^{x}$ is of the same form as the first term of $y_c$, then standard trial function $ae^{x}$ cannot be a particular integral* *as it automatically satisfies the homogenous equation $y''(x) - 3y'(x) + 2y(x) = 0$. We have to try $y_p^* = (a + bx)e^x$ first, and we have*

$$e^x(a + 2b + bx) - 3e^x(a + b + bx) + 2(a + bx)e^x = e^x.$$

*Dividing both sides by $e^x$, we have*

$$(a + 2b + bx) - 3(a + b + bx) + 2(a + bx) = 1,$$

*or*

$$b = -1.$$

*As there is no constraint for $a$, thus we take it to be zero ($a = 0$). In fact, any non-zero $ae^x$ can be absorbed into $Ae^x$. Thus, the general solution becomes*

$$y = -xe^x + Ae^x + Be^{2x}.$$

□

### 5.3.2 Differential Operator

A very useful technique is to use the method of differential operator $D$. A differential operator $D$ is defined as

$$D \equiv \frac{d}{dx}. \tag{5.17}$$

Since we know that $De^{\lambda x} = \lambda e^{\lambda x}$ and $D^n e^{\lambda x} = \lambda^n e^{\lambda x}$, so they are equivalent to $D \mapsto \lambda$, and $D^n \mapsto \lambda^n$. Thus, any polynomial $P(D)$ will map to $P(\lambda)$. On the other hand, the integral operator $D^{-1} = \int dx$ is just the inverse of the differentiation. The beauty of using the differential operator form is that one can factorize it in the same way as for factorizing polynomials, then solve each factor separately. Thus, differential operators are very useful in finding out both the complementary functions and particular integral.

□ ***Example 5.5:*** *To find the particular integral for the equation*

$$y''''' + 2y = 17e^{2x},$$

we get

$$(D^5 + 2)y^* = 17e^{2x}.$$

or

$$y^* = \frac{17}{D^5 + 2}e^{2x}.$$

Since $D^5 \mapsto \lambda^5 = 2^5$, we have

$$y^* = \frac{17e^{2x}}{2^5 + 2} = \frac{e^{2x}}{2}.$$

□

This method also works for $\sin x, \cos x, \sinh x$ and others, and this is because they are related to $e^{\lambda x}$ via $\sin\theta = \frac{1}{2i}(e^{i\theta} - e^{-i\theta})$ and $\cosh x = (e^x + e^{-x})/2$.

Higher order differential equations can conveniently be written as a system of differential equations. In fact, an $n$th-order linear equation can always be written as a linear system of $n$ first-order differential equations. A linear system of ODEs is more suitable for mathematical analysis and numerical integration.

## 5.4 Linear System

For a linear $n$ order equation (5.15), it can be always written as a linear system

$$\frac{dy}{dx} = y_1, \quad \frac{dy_1}{dx} = y_2, \quad ..., \quad \frac{dy_{n-1}}{dx} = y_{n-1},$$

$$a_n(x)y'_{n-1} = -a_{n-1}(x)y_{n-1} + ... + a_1(x)y_1 + a_0(x)y + \phi(x), \tag{5.18}$$

which is a system for $u = [y\ y_1\ y_2\ ...\ y_{n-1}]^T$.

For a second-order differential equation, we can always write it in the following form

$$\frac{du}{dx} = f(u, v, x), \qquad \frac{dv}{dx} = g(u, v, x). \tag{5.19}$$

If the independent variable $x$ does not appear explicitly in $f$ and $g$, then the system is said to be autonomous. Such a system has important properties. For simplicity and in keeping with the convention, we use $t = x$ and $\dot{u} = du/dt$ in our following discussion. A general linear system of $n$-th order can be written as

$$\begin{pmatrix} \dot{u}_1 \\ \dot{u}_2 \\ \vdots \\ \dot{u}_n \end{pmatrix} = \begin{pmatrix} a_{11} & a_{12} & \dots & a_{1n} \\ a_{21} & a_{22} & \dots & a_{2n} \\ \vdots & & & \vdots \\ a_{n1} & a_{n2} & \dots & a_{nn} \end{pmatrix} \begin{pmatrix} u_1 \\ u_2 \\ \vdots \\ u_n \end{pmatrix}, \tag{5.20}$$

or

$$\dot{\mathbf{u}} = \mathbf{A}\mathbf{u}. \tag{5.21}$$

If $\mathbf{u} = \mathbf{v}\exp(\lambda t)$, then this becomes an eigenvalue problem for matrix $\mathbf{A}$,

$$(\mathbf{A} - \lambda\mathbf{I})\mathbf{v} = \mathbf{0}, \tag{5.22}$$

which will have a non-trivial solution only if

$$\det(A - \lambda I) = 0. \tag{5.23}$$

## 5.5 Sturm-Liouville Equation

One of the commonly used second-order ordinary differential equation is the Sturm-Liouville equation in the interval $x \in [a, b]$

$$\frac{d}{dx}[p(x)\frac{dy}{dx}] + q(x)y + \lambda r(x)y = 0, \tag{5.24}$$

with the boundary conditions

$$y(a) + \alpha y'(a) = 0, \qquad y(b) + \beta y'(b) = 0, \tag{5.25}$$

where the known function $p(x)$ is differentiable, and the known functions $q(x), r(x)$ are continuous. The parameter $\lambda$ to be determined can only take certain values $\lambda_n$, called the eigenvalues, if the problem has solutions. For the obvious reason, this problem is called the Sturm-Liouville eigenvalue problem.

For each eigenvalue $\lambda_n$, there is a corresponding solution $\psi_{\lambda_n}$, called eigenfunctions. The Sturm-Liouville theory states that for two different eigenvalues $\lambda_m \neq \lambda_n$, their eigenfunctions are orthogonal. That is

$$\int_a^b \psi_{\lambda_m}(x)\psi_{\lambda_n}(x)r(x)dx = 0. \tag{5.26}$$

or more generally

$$\int_a^b \psi_{\lambda_m}(x)\psi_{\lambda_n}(x)r(x)dx = \delta_{mn}. \tag{5.27}$$

It is possible to arrange the eigenvalues in an increasing order

$$\lambda_1 < \lambda_2 < .... < \lambda_n < ... \to \infty. \tag{5.28}$$

Sometimes, it is possible to transform a nonlinear equation into a standard linear equation. For example, the Riccati equation can be written in the generic form

$$y' = p(x) + q(x)y + r(x)y^2, \qquad r(x) \neq 0. \tag{5.29}$$

If $r(x) = 0$, then it reduces to a first order linear ODE. By using the transform

$$y(x) = -\frac{u'(x)}{r(x)u(x)}, \tag{5.30}$$

or

$$u(x) = e^{-\int r(x)y(x)dx}, \tag{5.31}$$

we have

$$u'' - P(x)u' + Q(x)u = 0, \tag{5.32}$$

where

$$P(x) = -\frac{r'(x) + r(x)q(x)}{r(x)}, \qquad Q(x) = r(x)p(x). \tag{5.33}$$

### 5.5.1 Bessel Equation

The well-known Bessel equation

$$x^2y'' + xy' + (x^2 - \nu^2)y = 0, \tag{5.34}$$

is in fact an eigenvalue problem as it can be written as

$$(xy')' + (x - \frac{\nu^2}{x})y = 0. \tag{5.35}$$

Although $\nu$ can be any real values, but we only focus on the case when the values of $\nu$ are integers. In order to solve this equation, we assume that the solution can be written as a series expansion in the form

$$y(x) = x^s \sum_{n=0}^{\infty} a_n x^n = \sum_{n=0}^{\infty} a_n x^{n+s}, \qquad a_0 \neq 0, \tag{5.36}$$

where $s$ is a parameter to be determined. If $a_0 = 0$, we can always change the value of $s$, so that the first term of $a_n$ is not zero. Thus, we assume in general $a_0 \neq 0$. This method is often called the Frobenius method which is essentially an expansion in terms of a Laurant series. Thus, we have

$$\frac{dy}{dx} = \sum_{n=0}^{\infty} a_n (n+s) x^{n+s-1} \tag{5.37}$$

$$\frac{d^2y}{dx^2} = \sum_{n=0}^{\infty} a_n (n+s)(n+s-1) x^{n+s-2} \tag{5.38}$$

Substituting these expression into the Bessel equation, we have

$$\sum_{n=0}^{\infty} (k+s)(k+s-1) a_n x^{n+s} + \sum_{n=0}^{\infty} (n+s) a_n x^{n+s}$$

$$+ \sum_{n=0}^{\infty} a_n x^{n+s+2} - \nu^2 \sum_{n=0}^{\infty} a_n x^{n+s} = 0. \tag{5.39}$$

Equating the coefficients of the same power $x^n (n = 0, 1, 2, ...)$, we can get some recurrence relationships. For $n = 0$, we have

$$a_0(s^2 - \nu^2) = 0. \tag{5.40}$$

Since $a_0 \neq 0$, we thus have

$$s = \pm\nu. \tag{5.41}$$

From $n = 1$ terms, we have

$$a_1(2s + 1) = 0, \tag{5.42}$$

or

$$a_1 = 0, \quad (s \neq -\frac{1}{2}). \tag{5.43}$$

For the rest of terms $n = 2, 3, 4, ...,$ we have

$$a_n(n + s)(n + s - 1) + a_n(n + s) + a_{n-2} - \nu^2 a_n = 0, \tag{5.44}$$

or

$$a_n = -\frac{a_{n-2}}{(n + s)^2 - \nu^2} = -\frac{a_{n-2}}{n(n + 2\nu)}. \tag{5.45}$$

Since we now know that $a_1 = 0$, thus $a_3 = a_5 = a_7 = ... = a_1 = 0$. All the even terms contain the factor $a_0$, we finally have

$$y(x) = a_0 \sum_{k=0}^{\infty} \frac{(-1)^k \nu!}{2^{2k} k!(k + \nu)!} x^{2k+\nu} = a_0 J_\nu, \tag{5.46}$$

where we have used $n = 2k = 0, 2, 4, ...$ so that $k = 0, 1, 2, 3, ....$ The function $J_\nu$

$$J_\nu = \sum_{n=0}^{\infty} \frac{(-1)^n}{2^{2n} n!(n + \nu)! x^{2n+\nu}}, \tag{5.47}$$

is called the Bessel function of the order $\nu$. This is the Bessel function of the first kind. It has many interesting properties:

$$\frac{d}{dx}[x^\nu J_\nu(x)] = x^\nu J_{\nu-1}(x); \tag{5.48}$$

$$\sum_{\nu=-\infty}^{\infty} J_\nu = 1, \qquad J_{-\nu}(x) = (-1)^\nu J_\nu(x); \tag{5.49}$$

and

$$\int_0^x u J_0(u) du = x J_1(x). \tag{5.50}$$

There are other properties as well such as the orthogonality

$$\int_a^b x J_\nu(\alpha x) J_\nu(\beta x) dx = 0, \qquad (\alpha \neq \beta). \tag{5.51}$$

### 5.5.2 Euler Buckling

As an example, let us look at the buckling of an Euler column which is essentially an elastic rod with one pin-jointed end and the applied axial load $P$ at the other end. The column has a length of $L$. Its Young's modulus is $E$ and its second moment of area is $I = \int y^2 dA = const$ (for a given geometry). Let $u(x)$ be the transverse displacement, the Euler beam theory gives the following governing equation

$$\frac{EI}{P}\frac{d^2u}{dx^2} + u = 0, \tag{5.52}$$

or

$$u'' + \alpha^2 u = 0, \qquad \alpha^2 = \frac{P}{EI}, \tag{5.53}$$

which is an eigenvalue problem. Its general solution is

$$u = A \sin \alpha x + B \cos \alpha x. \tag{5.54}$$

Applying the boundary conditions, we have at the fixed end

$$u = 0 \qquad (\text{at } x = 0), \qquad B = 0, \tag{5.55}$$

and at the free end

$$u = 0, \qquad (\text{at } x = L), \qquad A \sin(\alpha L) = 0. \tag{5.56}$$

Thus we have two kinds of solutions either $A = 0$ or $\sin(\alpha L) = 0$. For $A = 0$, we have $u(x) = 0$ which is a trivial solution. So the non-trivial solution requires that

$$\sin(\alpha L) = 0, \tag{5.57}$$

or

$$\alpha L = 0 \text{ (trivial)}, \; \pi, \; 2\pi, \; ..., \; n\pi, \; ... \tag{5.58}$$

Therefore, we have

$$P = \alpha^2 EI = \frac{n^2\pi^2 EI}{L^2}, \qquad (n = 1, 2, 3, ...). \tag{5.59}$$

The solutions have fixed mode shapes (sine functions) at some critical values (eigenvalues $P_n$). The lowest eigenvalue is

$$P_* = \frac{\pi^2 EI}{L}, \tag{5.60}$$

which is the Euler buckling load for an elastic rod.

### 5.5.3 Nonlinear Second-Order ODEs

For higher-order nonlinear ordinary differential equations, there is no general solution technique. Even for relatively simple second-order ODEs, different equations will usually require different methods, and there is no guarantee that you can find the solution. One of the best methods is the change of variables so that the nonlinear equation can be transformed into a linear ordinary differential equation or one that can be solved by other methods. This can be beautifully demonstrated by the solution process of finding the orbit of a satellite.

As the satellite orbits the Earth, the force is purely radial in the polar coordinates, therefore, its total angular momentum $L$ is conserved. $L = mr^2\frac{d\theta}{dt} = const$, or $r^2\dot{\theta} = L/m = h = const$. The radial acceleration is $\mathbf{a}_r = \ddot{r} - r\dot{\theta}$. Using Newton's second law of motion and Newton's law of gravity, we have

$$m\left[\frac{d^2r}{dt^2} - r\left(\frac{d\theta}{dt}\right)^2\right] = -\frac{GMm}{r^2}, \tag{5.61}$$

where $M$ and $m$ are the masses of the Earth and the satellite, respectively. $G$ is the universal constant of gravitation. Using the conservation of angular momentum so that

$$r\dot{\theta}^2 = \frac{h^2}{r^3}, \qquad h = \frac{L}{m}, \tag{5.62}$$

we then have

$$\frac{d^2 r}{dt^2} - \frac{h^2}{r^3} + \frac{GM}{r^2} = 0, \tag{5.63}$$

which is a nonlinear equation. By using the change of variables $u = 1/r$, the conservation of angular momentum becomes

$$\frac{d\theta}{dt} = hu^2, \tag{5.64}$$

which is equivalent to $dt = d\theta/(hu^2)$ and this can be used to eliminate $t$. Then, we have

$$\frac{dr}{dt} = -u^{-2}\frac{du}{dt} = -h\frac{du}{d\theta}, \tag{5.65}$$

and

$$\frac{d^2 u}{dt^2} = \frac{d\dot{r}}{dt} = -h\frac{d^2 u}{d\theta^2}\frac{d\theta}{dt} = -h^2 u^2 \frac{d^2 u}{d\theta^2}. \tag{5.66}$$

Now the governing equation becomes

$$-h^2 u^2 \frac{d^2 u}{d\theta^2} - h^2 u^3 + GMu^2 = 0, \tag{5.67}$$

or

$$\frac{d^2 u}{d\theta^2} + u = \frac{GM}{h^2} \equiv S. \tag{5.68}$$

Since this is a second-order linear ordinary differential equation, it is straightforward to write down its solution

$$u = S + A\cos\theta + B\cos\theta = S[1 + e\cos(\theta + \psi)], \tag{5.69}$$

where $A$ and $B$ are integration constants, which can be converted into the eccentricity $e$ and the initial phase $\psi$. The final solution is

$$r = \frac{1}{S[1 + e\cos(\theta + \psi)]}, \tag{5.70}$$

which corresponds to an ellipse where $e$ is the eccentricity of the orbit. If we set the polar coordinates in such a way that $\psi = 0$ (say, along the major axis) and one focus at the origin, then the equation simply becomes

$$r = \frac{h^2}{GM[1 + e\cos\theta]}, \tag{5.71}$$

which is the orbit for satellites and planets.

# Chapter 6

# Recurrence Equations

## 6.1 Linear Difference Equations

Differential equations always concern the quantities that vary continuously. However, some problems such as finance are concerned with quantities (say, interest rate) that are discrete and do not vary continuously, and even the independent variables such as time are not continuously counted or measured (in seconds or years). For this type of problem, we need the difference equation or the recurrence equation as the counterpart in differential equations. In fact, there many similarity between difference equations and differential equations, especially the linear ones.

A linear difference equation of $N$-order can generally be written as

$$a_0 y_n + a_1 y_{n-1} + a_2 y_{n-2} + ... + a_N y_{n-N} = f(n), \qquad (6.1)$$

where $a_i (i = 0, ..., N)$ are coefficients which are not functions of $y$. $y_n$ $[\equiv y(n)]$ is the value of the variable $y$ at $n = 0, 1, 2, ....$ If $f(n) = 0$, we say that the difference equation is homogeneous. If all the coefficients $a_i$ are constant, the equation is called the linear equation with constant coefficients (as the counterpart in the differential equations). In this book, we only focus on the second-order linear difference equation. Thus, we have

$$ay_n + by_{n-1} + cy_{n-2} = f(n), \tag{6.2}$$

which can also be written as

$$ay_{n+1} + by_n + cy_{n-1} = g(n), \tag{6.3}$$

where $g(n) = f(n-1)$. If $f(n) = 0$, we say the equation is homogeneous. The most famous difference equation is probably the recurrence relation

$$y_n = y_{n-1} + y_{n-2}, \tag{6.4}$$

for the Fibonacci sequence $(0, 1, 1, 2, 3, 5, 8, ...)$. The recurrence equation is valid for $n = 2, 3, ...$ and the initial conditions are $y(0) = 0, y(1) = 1$.

Similar to the solution procedure of linear ordinary differential equations, the general solution of a difference equation $y_n = u_n + p_n$ where $u_n$ is the complementary solution to the homogeneous equation

$$ay_n + by_{n-1} + cy_{n-2} = 0, \tag{6.5}$$

while $p_n$ is any particular solution of (6.2).

In order to obtain $u_n$, we assume that $u_n = \alpha\lambda^n$ (similar to $y_c = Ae^{\lambda x}$ for differential equations). Substituting into (6.5), we reach a characteristic equation

$$a\lambda^2 + b\lambda + c = 0. \tag{6.6}$$

It has two solutions $\lambda_1$ and $\lambda_2$ in general. Therefore, we have

$$u_n = A\lambda_1^n + B\lambda_2^n. \tag{6.7}$$

---

□ ***Example 6.6:*** *Find the solution of*

$$y_n + 3y_{n-1} + 2y_{n-2} = 0.$$

*This is a homogeneous equation, we assume $u_n = \alpha\lambda^n$, we have*

$$\lambda^2 + 3\lambda + 2 = 0, \qquad \text{or} \qquad (\lambda + 1)(\lambda + 2) = 0.$$

*Thus, we have $\lambda_1 = -1$ and $\lambda_2 = -2$. Therefore, the general solution can be written as $y_n = A(-1)^n + B(-2)^n$.* □

For given initial values, we can determine the constant in the general solution so that an exact expression can be obtained.

---

□ ***Example 6.7:*** *The Fibonacci sequence is governed by the difference equation*

$$y_n = y_{n-1} + y_{n-2},$$

*with initial conditions $y_0 = 0, y_1 = 1$. This is a homogeneous equation. The characteristic equation is*

$$\lambda^2 - \lambda - 1 = 0,$$

*whose solution is $\lambda = \frac{1 \pm \sqrt{5}}{2}$. The general solution is therefore*

$$y_n = A(\frac{1+\sqrt{5}}{2})^n + B(\frac{1-\sqrt{5}}{2})^n.$$

*In order to determine $A$ and $B$, we first use $y_0 = 1$, we get*

$$0 = A + B.$$

*For $n = 1$, $y_1 = 1$ gives*

$$1 = A(\frac{1+\sqrt{5}}{2}) + B(\frac{1-\sqrt{5}}{2}).$$

*Now we have $A = 1/\sqrt{5}, B = -1/\sqrt{5}$. The general solution becomes*

$$y_n = \frac{1}{\sqrt{5}}(\frac{1+\sqrt{5}}{2})^n - \frac{1}{\sqrt{5}}(\frac{1-\sqrt{5}}{2})^n.$$

□

For finding a particular solution $p_n$, we can use the similar technique used in ordinary differential equations. For $f(n) = k = const$, we try $p_n = \alpha$. For $f(n) = kn$, we try $p_n = \alpha + \beta n$ where $\alpha$ and $\beta$ will be determined. For $f(n) = k\gamma^n$, we try

$p_n = \alpha\gamma^n$. Other forms such as polynomials can be done in a similar fashion.

---

□ ***Example 6.8:*** *For the equation*

$$y_{n+1} - y_n - 6y_{n-1} = n,$$

*its complementary equation is*

$$y_{n+1} - y_n - 6y_{n-1} = 0,$$

*and the characteristic equation is*

$$\lambda^2 - \lambda - 6 = 0, \qquad \text{or} \qquad (\lambda + 2)(\lambda - 3) = 0.$$

*The general complementary solution can now be written as*

$$u(n) = A(-2)^n + B3^n.$$

*Finding any particular solution requires that we try $p_n = a + bn$. We now have*

$$a + b(n+1) - (a + bn) - 6[a + b(n-1)] = n,$$

*or*

$$-6a + 7b - 6bn = n.$$

*As for any $n$, this equation is valid, therefore we have $a = \frac{7}{36}$, and $b = -\frac{1}{6}$. Finally, the general solution is*

$$y(n) = A(-2)^n + B3^n - \frac{n}{6} - \frac{7}{36}.$$

□

---

## 6.2 Chaos and Dynamical Systems

As you may have noticed that the above analysis is mainly about the linear equations, what happens to the nonlinear equations? The main problem with nonlinear equations is that there is no general solution techniques available for most cases. Even for the simplest case, the analysis is not easy. In addition, the behaviour of nonlinear equations is very complex, even for the simplest equations. Often, nonlinear equations may have chaotic behaviour under appropriate conditions.

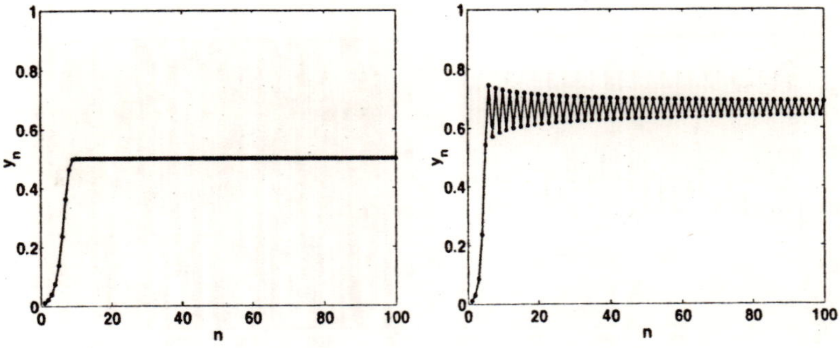


Figure 6.1: Variations of $y_n$ for $\nu = 2$ and $\nu = 3$.

### 6.2.1 Bifurcations and Chaos

In order to show this, we now briefly introduce the concept of chaos by studying the following nonlinear difference equation

$$y_{n+1} = \nu y_n(1 - y_n), \tag{6.8}$$

where $\nu$ is a constant parameter. This is the well-studied logistic map, which is essentially an iteration process because all the values $y_2, y_3, \ldots$ can be determined for a given parameter $\nu$ and an initial condition $y_1$. This is one of the simplest dynamical systems. This seemingly simple equation is in fact very complicated. If you try to use the method to solve the linear difference equations discussed earlier, it does not work.

For a given value $\nu = 2$, we can use a computer or a pocket calculator to do these calculations. If the initial value $y_1 = 0$ or $y_1 = 1$, then the system seems to be trapped in the state $y_n = 0$ (n=2,3, ...). However, if we use a slight difference value (say) $y_1 = 0.01$, then we have

$$y_1 = 0.01, y_2 = 0.0198, y_3 = 0.0388, y_4 = 0.0746,$$

$$y_5 = 0.1381, y_6 = 0.2381, y_7 = 0.3628, y_8 = 0.4623,$$

$$y_9 = 0.4972, \quad y_{10} = 0.5000, \quad y_{11} = 0.5000, \quad \ldots \tag{6.9}$$

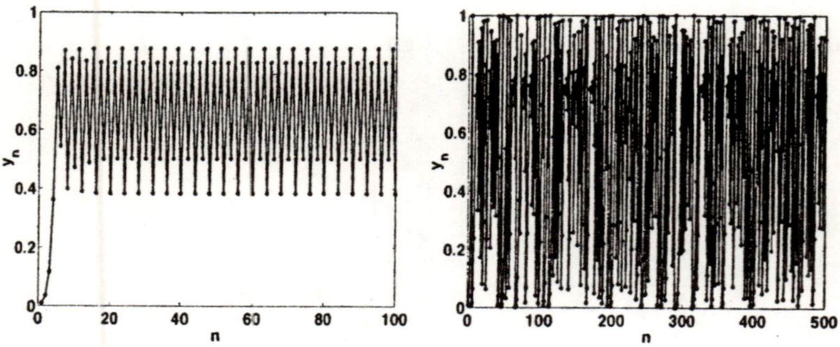


Figure 6.2: Variations of $y_n$ for $\nu = 3.5$ and $\nu = 4$.

Then, the values are trapped or attracted to a single value or state $y_\infty = 0.5000$. All the values are then plotted in a graph as shown in Figure 6.1.

If we use a different parameter $\nu = 3$ and run the simulations again from the same initial value $y_1 = 0.01$, the results are also plotted on the right in Figure 6.1. Now we have a difference phenomenon. The final values do not settle to a single value. Instead, they oscillate between two states or two final values $y_\infty = 0.6770$ and $y_\infty * = 0.6560$. The iteration system bifurcates into two states as the parameter $\nu$ increases. If we do the same simulations again using a different value $\nu = 3.5$ (shown in Figure 6.2), there are four final states. For $\nu = 4$, every values seems difference, the system is chaotic and the values looks like a random noise.

Following exactly the same process but using different values of $\nu$ ranging from $\nu = 0.001$ to $\nu = 4.000$, we can plot out the number of states (after $N = 500$ iterations) and then we get a bifurcation map shown in Figure 6.3. It gives a detailed map about how the system behaves. From the chaotic map, we see that for $\nu < 1$, the final state is zero (the system is attracted to a stable state $y_\infty = 0$). For $1 < \nu < 3$, the system settles (or attracts) to a single state. For $3 < \nu < 3.45$, the system bifurcates into two states. It seems that the system is attracted

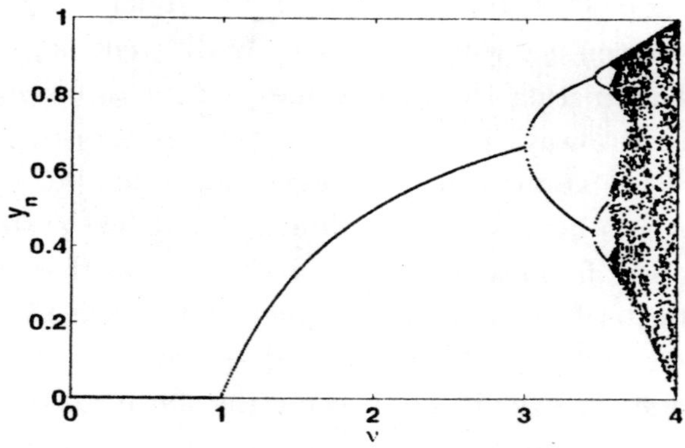


Figure 6.3: Bifurcation diagram and chaos.

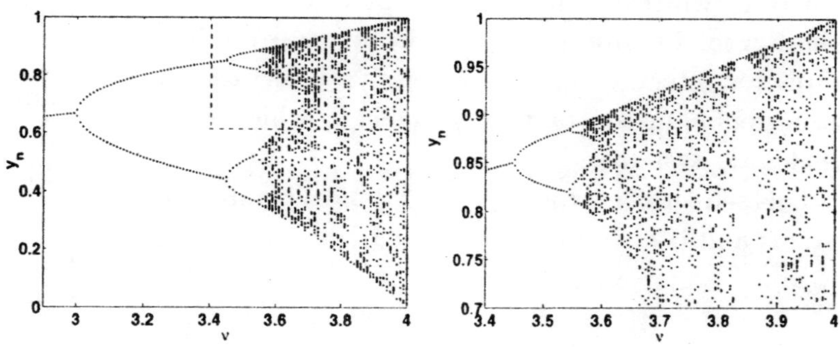


Figure 6.4: Bifurcation and similarity.

by these discrete states. For this reason, the map is also called the attractors of the dynamical system. The system becomes chaotic for $\nu \geq \nu_*$ where $\nu_* \approx 3.57$ is the critical value.

It is a bit surprising for a seemingly determined system $\nu y_n(1 - y_n) \to y_{n+1}$ because you may try many times to simulate the same system at the same initial value $y_0 = 0.01$ (say) and parameter $\nu$. Then, we should get the same set of values $y_1, y_2, \ldots$. You are right. So where is the chaos anyway? The problem is that this system is very sensitive to the small variation in the initial value $y_0$. If there is a tiny different, say,

$y_0 = 0.01 \pm 0.000000001$ or even $10^{-1000}$ difference, then the set of values you get will be completely different, and there is no way of predicting the final values (say, $y_{500}$ or $y_{1,000,000}$). Since there is always uncertainty in the real world, even the computer simulations can only use a finite number of digits, so the chaos is intrinsic for nonlinear dynamical systems. In fact, there is a famous 'butterfly effect'. It says that the wing flip of a butterfly in Africa can cause a tornado in America or anywhere. Obviously, this is exaggerated too much, but it does provide some vivid picture for the nature of chaos and sensitivity to the initial uncertainty in chaotic systems.

If we study Figure 6.4 closely, we can see there is a similarity between the whole map and its certain part (enclosed by a box) which is enlarged and plotted on the right in the same figure. In addition, the ratio between the lengths of the parameter intervals for two successive bifurcation approaches the Feigenbaum constant $\delta_F = 4.669...$, which is universal for the chaotic systems.

This self-similarity is one of the typical behaviours of chaotic systems and it also occurs in other nonlinear systems such as

$$y_{n+1} = \lambda \sin y_n, \tag{6.10}$$

and

$$y_{n+1} = \lambda \sin^2 y_n, \tag{6.11}$$

which are plotted in Figure 6.5.

### 6.2.2 Dynamic Reconstruction

When a nonlinear system becomes chaotic, it seems that it is very difficult to understand the behaviour. However, there may be some regularity such as attractors and self-similarity as we have seen earlier. In some case, it is even possible to reconstruct the system itself.

Suppose we do not know the exact form of the dynamical system, but we do know it only depends on one parameter

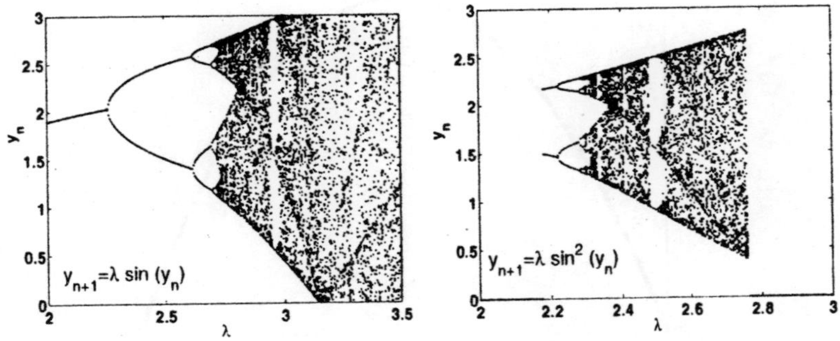


Figure 6.5: Bifurcation for $y_{n+1} = \lambda \sin y_n$ and $y_{n+1} = \lambda \sin^2 y_n$.

and we can observe the states $y_1, y_2, ..., y_n$ (to a certain degree). From our observations, we can actually reconstruct the system using the sorted data and plotting $y_n$ versus $y_{n-1}$. If there are $N$ observations, we have $U = y_2, ..., y_N$ as one set and $V = y_1, y_2, ..., y_{n-1}$ as another set. We plot $U$ versus $V$, then the system can be dynamically reconstructed. For example, from the 100 data for the nonlinear system discussed in the previous section, the constructed system is a parabola as plotted in Figure 6.6. The parabola is essentially the original function $y(1-y)$. The mapping is then $y \mapsto y(1-y)$. With a free parameter $\nu$ and discrete time $n$, we obtain the dynamical system

$$y_n = \nu y_{n-1}(1 - y_{n-1}), \qquad \text{or} \qquad y_{n+1} = \nu y_n(1 - y_n). \quad (6.12)$$

We see that even a simple nonlinear equation in 1-D can show the rich complexity of dynamical behaviour. Now we briefly look at a Lorenz system as a classical example.

### 6.2.3 Lorenz Attractor

The Lorenz attractor was first discovered by Edward Lorenz when he studied the weather model in 1963. The Lorenz equa-

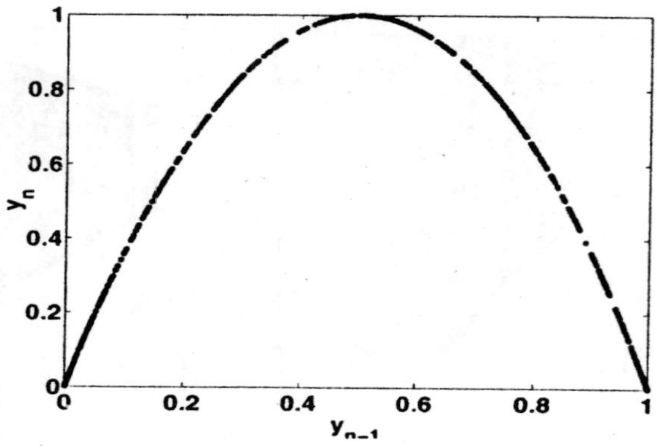


Figure 6.6: Dynamic reconstruction of the nonlinear function.

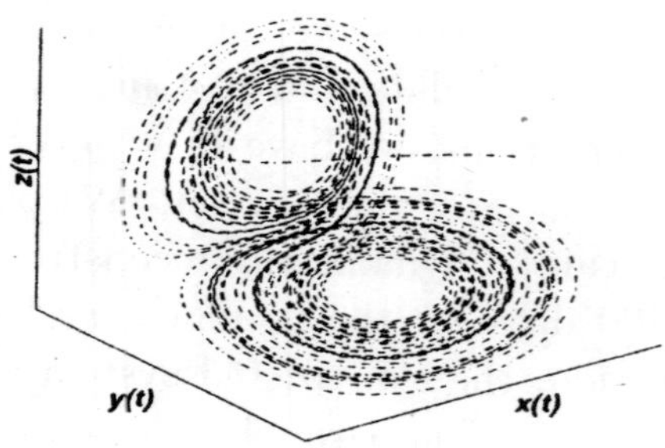


Figure 6.7: Lorenz strange attractor.

tions are

$$\dot{x} = -\beta x + yz,$$
$$\dot{y} = \delta(z - y),$$
$$\dot{z} = (\gamma - x)y - z, \tag{6.13}$$

where $\dot{} = \frac{d}{dt}$, and $\beta, \delta, \gamma$ are parameters. Specifically, $\delta$ is the Prandtl number, and $\gamma$ is the Rayleigh number.

For certain ranges of parameters, for example $\beta = 8/3, \delta = 10, \gamma = 28$, the system becomes chaotic. The system moves around in a curious orbit in 3-D as shown in Figure 6.7, and the orbit is bounded, but not periodic or convergent. This strange characteristic gives the name *'strange'* to the attractor, and thus the Lorenz strange attractor. The chaotic map looks like a butterfly.

## 6.3 Self-similarity and Fractals

In the bifurcation map of chaotic systems, we have seen the self-similarity. This self-similarity is quite universal in many phenomena, and it is also linked with the concept of fractals. In fact, self-similarity and fractals occur frequently in nature. Two classical examples are fern leaves and the pattern of lightening. This observation suggests that even if the system is chaotic, there is still some pattern or certain regularity in its behaviour, and thus chaos is very different from the random noise, even though they sometimes may look the same.

In geometry, we know that a point is zero dimension and a line is one dimension. Similarly, the dimension for a plane is two and the dimension for a volume is three. All these dimensions are integers $(0, 1, 2, 3)$. But there are other possibilities in nature that the dimension can be a fraction or a real number. This is where the fractals arise. The basic idea of fractals can be demonstrated using the generation of the Koch curve (also called the Koch snowflake curve). The generator or seed of this curve is the four straight line segments as shown in Figure 6.8. By starting with this generator (left) denoted as $S_1$, and replacing each straight line segment by the generator itself, we get the second generation curve (middle, $S_2$). If we continue this procedure many times, we get the Koch curve. A snapshot at generation 5 is shown on the right in Figure 6.8. The total length of the seed $S_1$ is $L_1 = 4/3$ since it has four segments and each segment has a length of $1/3$, for the $S_2$ curve, the total length is $4^2/3^2$. This is because there are 4 more segments added, but each segment is only $1/3$ of the previous one in length. For the $n$-th generation, we have the total length

Figure 6.8: Generation of the Koch curve.

$(\frac{4}{3})^n$ with $4^n$ segments in total.

The fractal dimension can be defined by covering the curve using small balls (circles in a plane) and counting the number of balls with radius $r = \epsilon > 0$. For a curve $\Gamma$ in a compact set of a metric space and for each $\epsilon$ as the radius of the small balls, the smallest number $N(\epsilon)$ of balls to cover the curve $\Gamma$ will varies with $\epsilon$. The fractal dimension is defined by the limit

$$d = -\lim_{\epsilon \to 0} \frac{\ln N(\epsilon)}{\ln \epsilon}. \tag{6.14}$$

For the Koch curve, the minimum radius of the balls is $\epsilon = 1/3^n$, and the total number is $N(\epsilon) = 4^n$. Hence, we have the fractal dimension of the Koch curve

$$d = -\lim_{\epsilon \to 0} \frac{\ln 4^n}{\ln 3^n} = \frac{\ln 4}{\ln 3} \approx 1.2619. \tag{6.15}$$

Another famous example of fractals, which is related to the iterative dynamical systems, is the Mandelbrot set $z = z^2 + c$, or

$$z_{n+1} = z_n^2 + c, \tag{6.16}$$

where $z \in C$ is in the complex plane, and $c$ is a complex number to be tested. If you start with $z_0 = 0$ (say), the iterations continue until certain criteria are met. In this set, the stopping criterion is simply $|z| \geq 2$. That is to say, for each given $c$, the magnitude of $z$ during the iterations changes, it either stays small ($< 2$) or it will eventually surpass two ($> 2$). If $|z| < 2$, we say that $c$ belongs to the Mandelbrot set, otherwise, it is not

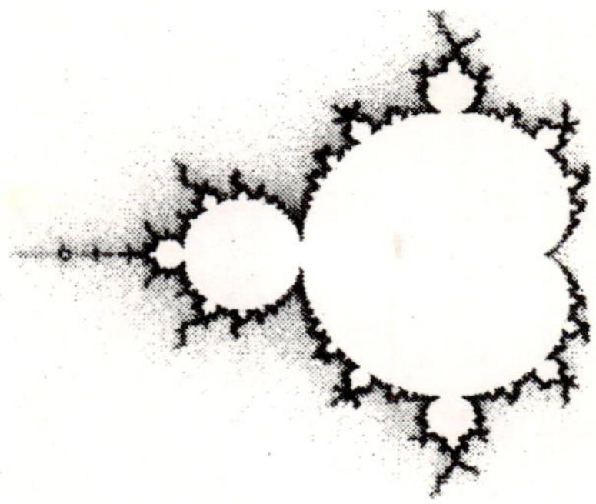

Figure 6.9: Fractal nature of the Mandelbrot set: $z = z^2 + c$.

part of the Mandelbrot set. The iterations will go on until the modulus of $z$ reach 2, and the point $c$ is marked on the complex plane $(x, iy)$ if it is not part of the Mandelbrot set. Then, we change a different value of $c$, and follow the same iterative process again. After many iterations across a region of the complex plane, the results become the well-known picture of the Mandelbrot set (shown in Figure 6.9). It is an usual practice to mark the points with colours depending on the number of iterations for each point to reach the modulus $|z| = 2$. This simple iterative system can produce beautiful patterns. You can view this system as a dynamical system, but it is a very complicated system.

Vast literature exists on this subject, and it is still an active research area.

# Chapter 7

# Vibration and Harmonic Motion

After the studies of the complex numbers and ordinary differential equations, it is time to see how they are applied to engineering problems. This chapter concerns the vibration and harmonic motion of mechanical systems.

## 7.1 Undamped Forced Oscillations

The simple system with a spring attached with a mass $m$ is a good example of harmonic motion (see Figure 7.1). If the spring stiffness constant is $k$, then the governing equation of the oscillations is a second-order ordinary differential equation for undamped forced harmonic motion, which can be written as

$$y'' + \omega_0^2 y = f(t). \tag{7.1}$$

where $\omega_0^2 = k/m$, and $f(t)$ is the a known function of $t$. In the case of $f(t) = \alpha \cos \omega t$, we have

$$y'' + \omega_0^2 y = \alpha \cos \omega t, \tag{7.2}$$

where $\omega_0$ is the natural frequency of the system, and $\alpha$ is the amplitude of external forcing.

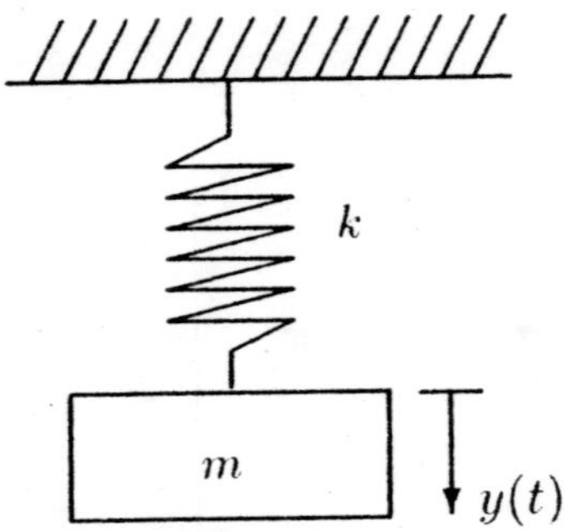

Figure 7.1: Undamped forced oscillations.

The general solution $y(t) = y_c + y_p$ consists of a complementary function $y_c$ and a particular integral $y_p$. The complementary function $y_c$ satisfies the homogeneous equation

$$y'' + \omega_0^2 y = 0. \tag{7.3}$$

Its general solution is

$$y_c(t) = A \sin \omega_0 t + B \cos \omega_0 t. \tag{7.4}$$

For the particular integral $y_p$, we have to consider two different cases $\omega \neq w_0$ and $\omega = \omega_0$ because for $\omega = \omega_0$ the standard particular $a \sin \omega t + b \cos \omega t$ does not work. It needs some modifications. For $\omega \neq \omega_0$, we assume that $y_p = a \sin \omega t + b \cos \omega t$. We thus obtain

$$y_p = \frac{\alpha}{\omega_0^2 - \omega^2} \cos \omega t. \tag{7.5}$$

Therefore, the general solution

$$y(t) = A \sin \omega_0 t + B \cos w_0 t + \frac{\alpha}{\omega_0^2 - \omega^2} \cos \omega t. \tag{7.6}$$

If we further assume that the system is initially at rest when the force starts to act, we have the initial conditions $y(0) = 0$ and $y'(0) = 0$. With these conditions, we have $A = 0$ and $B = -\alpha/(\omega_0^2 - \omega^2)$ in the general solution. We now have

$$y(t) = \frac{\alpha}{\omega_0^2 - \omega^2} (\cos \omega t - \cos \omega_0 t). \tag{7.7}$$

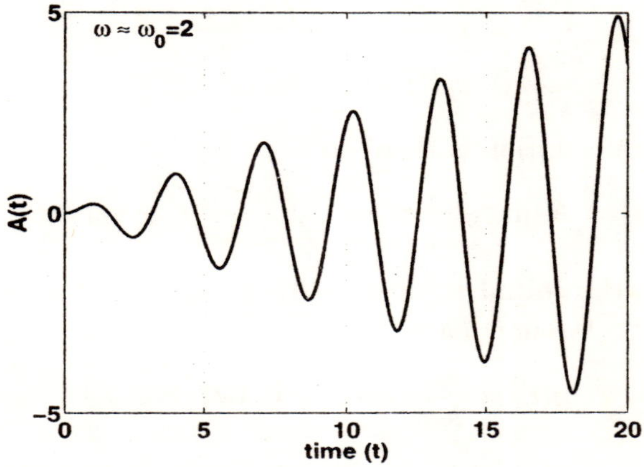


Figure 7.2: Variation of amplitude at $\omega = \omega_0$.

Using

$$\cos C - \cos D = -2 \sin \frac{(C+D)}{2} \sin \frac{(C-D)}{2},$$

we have

$$y(t) = \frac{2\alpha}{\omega_0^2 - \omega^2} \sin \frac{(\omega - \omega_0)t}{2} \sin \frac{(\omega + \omega_0)t}{2}$$

$$= A(t) \sin \frac{(\omega + \omega_0)t}{2} = A(t) \sin \tilde{\omega} t, \tag{7.8}$$

where

$$A(t) = \frac{2\alpha}{\omega_0^2 - \omega^2} \sin \frac{(\omega - \omega_0)t}{2}. \tag{7.9}$$

As $|\omega - \omega_0| < |\omega + \omega_0|$, we can see that the oscillator oscillates with a major and fast frequency $\tilde{\omega} = (\omega + \omega_0)/2$, while its amplitude or envelope slow oscillates with a frequent $\delta\omega = (\omega - \omega_0)/2$. This phenomenon is called beats.

For the special case of $\omega = \omega_0$, the complementary function is the same as before, but the particular solution should take the following form

$$y_p = t(a \sin \omega t + b \cos \omega t), \tag{7.10}$$

which gives

$$y_p(t) = \frac{\alpha}{2\omega_0} t \sin \omega_0 t. \tag{7.11}$$

The general solution is therefore

$$y(t) = A \sin \omega_0 t + B \cos \omega_0 t + \frac{\alpha}{2\omega_0} t \sin \omega_0 t. \tag{7.12}$$

Similarly, the initial solution $y(0) = y'(0) = 0$ implies that $A = B = 0$. We now have

$$y(t) = \frac{\alpha}{2\omega_0} t \sin \omega_0 t = A(t) \sin \omega_0 t, \tag{7.13}$$

where $A(t) = \alpha t/(2\omega_0)$. As the amplitude $A(t)$ increases with time as shown in Figure 7.5, this phenomenon is called resonance, and the external forcing cause the oscillations grow out of control when the forcing is acted at the natural frequency $\omega_0$ of the system.

## 7.2 Damped Forced Oscillations

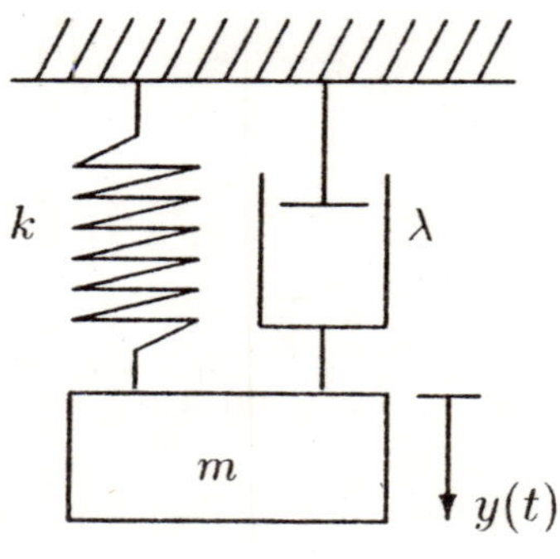


Figure 7.3: Damped harmonic motion.

As all the real systems have a certain degree of friction, thus damping should be included. An example of damping is shown in Figure 7.3. With damping, the equation of forced oscillations becomes

$$y''(t) + 2\lambda y'(t) + \omega_0^2 y(t) = \alpha \cos \omega t, \tag{7.14}$$

where $\lambda$ is the damping coefficient. In principle, one can try to solve this equation using the standard method, but it may become a little awkward as it involves complex numbers. In fact, there is quite an elegant method using the complex variables. In order to do this, we write the companion equation for equation (7.14) with a different forcing term

$$\eta''(t) + 2\lambda\eta'(t) + \omega_0^2\eta(t) = \alpha \sin \omega t. \tag{7.15}$$

Since $e^{i\omega t} = \cos \omega t + i \sin \omega t$, we can multiply (7.15) by $i$, and add it to (7.14), and we have

$$z''(t) + 2\lambda z + \omega_0^2 z = \alpha e^{i\omega t}, \tag{7.16}$$

where $z(t) = y(t) + i\eta(t)$. By solving this equation, we essentially solve both equations (7.14) and (7.15) at the same time if we can separate the real and imaginary parts. The complementary function corresponds to the transient part while the particular function corresponds to the steady state. For the transient part, the characteristic equation gives

$$\mu^2 + 2\lambda\mu + \omega_0^2 = 0, \tag{7.17}$$

or

$$\mu = -\lambda \pm \sqrt{\lambda^2 - \omega_0^2}. \tag{7.18}$$

If $\lambda^2 \geq \omega_0^2$, then $\mu < 0$. If $\lambda^2 < \omega_0^2$, then $\mu = -\lambda + i\sqrt{\omega_0^2 - \lambda^2}$ and $\Re(\mu) < 0$. In both cases $\Re(\mu) < 0$, thus the solution $z_c \propto e^{-\Re(\mu)t} \to 0$. In engineering, it is conventional to define a case of critical damping when $\xi = \lambda/\omega_0 = 1$. The quality factor $Q = \frac{1}{2\xi}$ is also commonly used. We now have

$$\mu = \omega_0(-\xi \pm \sqrt{\xi^2 - 1}). \tag{7.19}$$

For $\xi = 0$, we have $\mu = i\omega_0$, which corresponds to the harmonic oscillations without damping. For $\xi = 1$, $\mu = -\omega_0$, it is critical damping as the imaginary term is zero. The amplitude decreases exponentially at just the slowest possible manner without any oscillations. For $\xi < 1$, we get $\mu = -\omega_0\xi + i\omega_0\sqrt{1 - \xi^2}$.

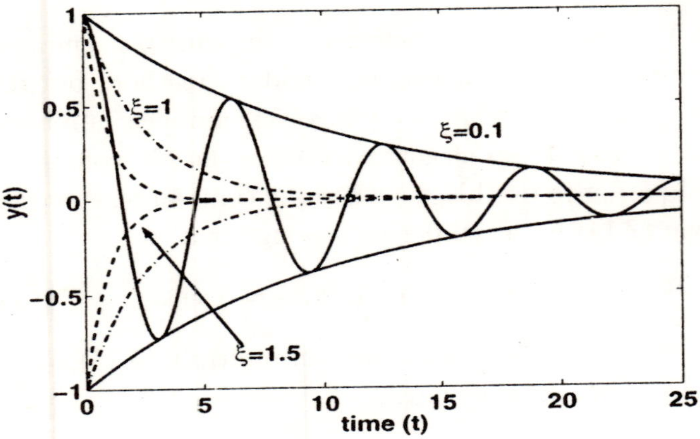


Figure 7.4: Critical damping ($\xi = 1$), under-damping ($\xi = 0.1$), and over-damping ($\xi = 1.5$).

The real part corresponds to the exponential decrease of the amplitude and the imaginary part corresponds to oscillations. For this reason, it is called under-damped. Finally, $\xi > 1$ leads to $\mu = \omega_0(-\xi \pm \sqrt{\xi^2 - 1}) < 0$. The imaginary part is zero (no oscillation). As the amplitude decreases much faster than that at the critical damping, this case is thus called over-damped. Figure 7.4 shows the characteristics of these three cases.

If time is long enough ($t \gg 1$), the transient part $y_c$ will becomes negligible as $t$ increases. Therefore, we only need to find the particular solution $z_p$.

If we try the particular solution in the form $z = z_0 e^{i\omega t}$, we have

$$z'' + 2\lambda z' + \omega_0^2 = P(i\omega)z, \tag{7.20}$$

and

$$P(i\omega) = (i\omega)^2 + 2\lambda(i\omega) + \omega_0^2 = (\omega_0^2 - \omega^2) + 2\lambda\omega i, \tag{7.21}$$

which is essentially the characteristic polynomial. The general solution becomes

$$z(t) = \frac{\alpha}{P(i\omega)} e^{i\omega t} = \frac{\alpha}{[(\omega_0^2 - \omega^2) + 2i\lambda\omega]} e^{i\omega t}. \tag{7.22}$$

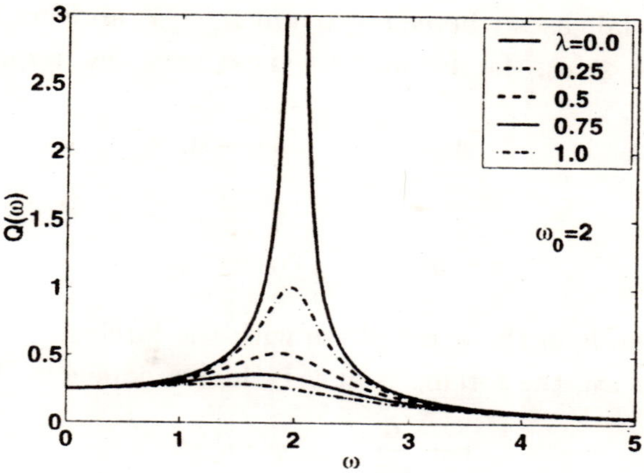


Figure 7.5: Variations of amplitude with frequency $\omega$ and damping coefficient $\lambda$.

It is conventional to call $H(i\omega) = 1/P(i\omega)$ the transfer function. We can always write the general solution $z = Ae^{i(\omega t+\phi)}$, where $A = |z|$ is the modulus and $\phi$ is the phase shift. Therefore, we have

$$z = Ae^{i(\omega t+\phi)}, \tag{7.23}$$

where

$$A = \frac{\alpha}{\sqrt{(\omega_0^2 - \omega^2)^2 + 4\lambda^2\omega^2}}, \tag{7.24}$$

and

$$\phi = \tan^{-1} \frac{-2\lambda\omega}{\omega_0^2 - \omega^2}. \tag{7.25}$$

As the amplitude of the forcing is $\alpha$, the gain $G(\omega)$ of the oscillation is

$$G(\omega) = \frac{A}{\alpha} = \frac{1}{\sqrt{(\omega_0^2 - \omega^2)^2 + 4\lambda^2\omega^2}}, \tag{7.26}$$

which is shown in Figure 7.5.

Finally, the solution of the original equation (7.14) is the real part. That is

$$y(t) = A\cos(\omega t + \phi). \tag{7.27}$$

Some special cases where $\omega \to 0$ and $\omega \to \infty$ are very interesting. For $\omega \ll \omega_0$, the driving force is at very low frequency, we have

$$A \to \frac{\alpha}{\omega_0^2}, \qquad \phi \to 0. \tag{7.28}$$

That is

$$y(t) \approx \frac{\alpha}{\omega_0^2} \cos wt. \tag{7.29}$$

The system is in the same phase with the forcing.

If $\omega \gg \omega_0$, the forcing is at very high frequency. We have

$$A \to \frac{\alpha}{\omega_0^2}, \qquad \phi \to -\pi. \tag{7.30}$$

The oscillator is completely out of phase with the forcing.

If $\omega \approx \omega_0$, we have

$$A \to \frac{\alpha}{2\lambda\omega_0}, \qquad \phi \to -\frac{\pi}{2}, \tag{7.31}$$

and

$$y(t) = \frac{\alpha}{2\lambda\omega_0} \sin \omega_0 t. \tag{7.32}$$

At resonance frequency $\omega_*^2 = \omega_0^2 - 2\lambda^2$, the amplitude of the oscillations increases dramatically.

## 7.3 Normal Modes

In the above harmonic oscillations, we know $\omega_0$ is the natural frequency of the concerned system, which is relatively simple. In general, there may be many natural frequencies or modes in a system, and the natural frequencies are in fact determined from the eigenvalue problem resulting from the system. Now let us study a more complicated system with three mass blocks attached in by two springs as shown in Figure 7.6. This system can be thought of as a car attached to two caravans on a flat road.

Let $u_1, u_2, u_3$ be the displacement of the three mass blocks $m_1, m_2, m_3$, respectively. Then, their accelerations will be $\ddot{u}_1$,

$\ddot{u}_2$, $\ddot{u}_3$ where $\ddot{u} = d^2u/dt^2$. From the balance of forces and Newton's law, we have

$$m_1\ddot{u}_1 = k_1(u_2 - u_1), \tag{7.33}$$

$$m_2\ddot{u}_2 = k_2(u_3 - u_2) - k_1(u_2 - u_1), \tag{7.34}$$

$$m_3\ddot{u}_3 = -k_2(u_3 - u_2). \tag{7.35}$$

These equations can be written in a matrix form as

$$\begin{pmatrix} m_1 & 0 & 0 \\ 0 & m_2 & 0 \\ 0 & 0 & m_3 \end{pmatrix} \begin{pmatrix} \ddot{u}_1 \\ \ddot{u}_2 \\ \ddot{u}_3 \end{pmatrix}$$

$$+ \begin{pmatrix} k_1 & -k_1 & 0 \\ -k_1 & k_1 + k_2 & -k_2 \\ 0 & -k_2 & k_2 \end{pmatrix} \begin{pmatrix} u_1 \\ u_2 \\ u_3 \end{pmatrix} = \begin{pmatrix} 0 \\ 0 \\ 0 \end{pmatrix}, \tag{7.36}$$

or

$$\mathbf{M}\ddot{\mathbf{u}} + \mathbf{K}\mathbf{u} = \mathbf{0}, \tag{7.37}$$

where $\mathbf{u}^T = (u_1, u_2, u_3)$. The mass matrix is

$$\mathbf{M} = \begin{pmatrix} m_1 & 0 & 0 \\ 0 & m_2 & 0 \\ 0 & 0 & m_3 \end{pmatrix}, \tag{7.38}$$

and the stiffness matrix is

$$\mathbf{K} = \begin{pmatrix} k_1 & -k_1 & 0 \\ -k_1 & k_1 + k_2 & -k_2 \\ 0 & -k_2 & k_2 \end{pmatrix}. \tag{7.39}$$

Equation (7.37) is a second-order ordinary differential equation in terms of matrices. This homogeneous equation can be solved by substituting $u_i = U_i \cos(\omega t)$ where $U_i (i = 1, 2, 3)$ are constants and $\omega^2$ can have several values which correspond to the natural frequencies. Now we have

$$-\omega_i^2 \mathbf{M} U_i \cos(\omega t) + \mathbf{K} U_i \cos(\omega t) = 0. \tag{7.40}$$

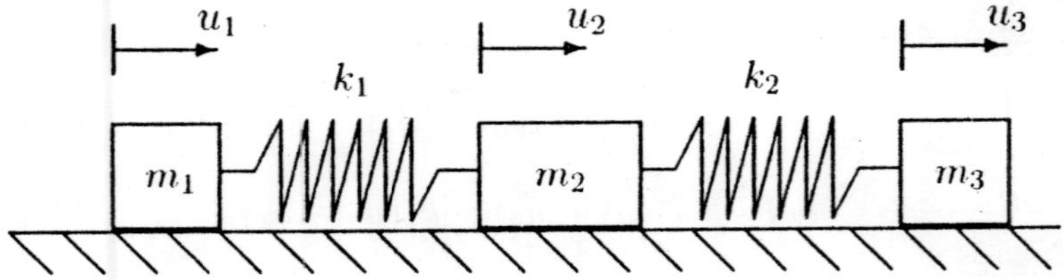


Figure 7.6: Harmonic vibrations.

where $i = 1, 2, 3$. Dividing both sides by $\cos(\omega t)$, we have

$$[\mathbf{K} - \omega^2 \mathbf{M}]U_i = 0. \tag{7.41}$$

This is essentially an eigenvalue problem because any non-trivial solutions for $U_i$ require

$$|\mathbf{K} - \omega^2 \mathbf{M}| = 0. \tag{7.42}$$

Therefore, the eigenvalues of this equation give the natural frequencies.

---

□ ***Example 7.1:*** *For the simplest case when $m_1 = m_2 = m_3 = m$ and $k_1 = k_2 = k$, we have*

$$\begin{vmatrix} k-\omega^2 m & -k & 0 \\ -k & 2k-\omega^2 m & -k \\ 0 & -k & k-\omega^2 m \end{vmatrix} = 0,$$

*or*

$$-\omega^2(k - \omega^2 m)(3km - \omega^2 m^2) = 0.$$

*This is a cubic equation in terms of $\omega^2$, and it has three solutions. Therefore, the three natural frequencies are*

$$\omega_1^2 = 0, \qquad \omega_2^2 = \frac{k}{m}, \qquad \omega_3^2 = \frac{3k}{m}.$$

*For $\omega_1^2 = 0$, we have $(U_1, U_2, U_3) = \frac{1}{\sqrt{3}}(1, 1, 1)$, which is the rigid body motion. For $\omega_2 = k/m$, the eigenvector is determined by*

$$\begin{pmatrix} 0 & -k & 0 \\ -k & k & -k \\ 0 & -k & 0 \end{pmatrix} \begin{pmatrix} U_1 \\ U_2 \\ U_3 \end{pmatrix} = \begin{pmatrix} 0 \\ 0 \\ 0 \end{pmatrix},$$

*which leads to $U_2 = 0$, and $U_1 = U_3$. Written in normalized form, it becomes $(U_1, U_2, U_3) = \frac{1}{\sqrt{2}}(1, 0, -1)$. This means that block 1 moves in the opposite direction away from block 3, and block 2 remains stationary. For $\omega_3^2 = 3k/m$, we have $(U_1, U_2, U_3) = \frac{1}{\sqrt{6}}(1, -2, 1)$. That is to say, block 2 moves in the different direction from block 3 which is at the same pace with block 1.* □

## 7.4 Small Amplitude Oscillations

For a mechanically conservative system, its total energy $E = T + V$ is conserved, where $T$ is its total kinetic energy and $V$ is its total potential energy. The configuration of the mechanical system can be described by its general coordinates $\mathbf{q} = (q_1, q_2, ..., q_n)$. The general coordinates can be distance and angles. Thus, the velocities of the system will be $\dot{\mathbf{q}} = \dot{q}_1, \dot{q}_2, ..., \dot{q}_n$. If we consider the system consists of many small particles or even imaginary parts, then the total kinetic energy $T$ is a function of $\dot{\mathbf{q}}$ and sometimes $\mathbf{q}$, but the potential energy $V$ is mainly a function of $\mathbf{q}$ only. As we are only concerned with small amplitude oscillations near equilibrium $V_{\min} = V(0) = V_0$, we can always take $\mathbf{q} = 0$ at the equilibrium so that we can expand $V$ in terms of $\mathbf{q}$ as a Taylor series

$$V(\mathbf{q}) = V_{min} + \sum_i \frac{\partial V_0}{\partial q_i} q_i + \sum_i \sum_j K_{ij} q_i q_j + ..., \tag{7.43}$$

where the stiffness matrix is

$$K_{ij} = \frac{1}{2} \frac{\partial^2 V_0}{\partial q_i \partial q_j}\Big|_{q_i=0, q_j=0}. \tag{7.44}$$

Since potential energy is always relative to an arbitrary reference point, we can thus take the potential energy at equilibrium $V_{min}$ to be zero. In addition, the equilibrium or the minimum value of $V$ requires $\frac{\partial V}{\partial q_i} = 0$ at the equilibrium point $q_i = 0$, and the force $F_i = \frac{\partial V}{\partial q_i}$ shall be zero. This is correct because the resultant force must be zero at equilibrium, otherwise, the system will be driven away by the resultant force. The component of the resultant force along the general coordinate $q_i$ should also be zero. Therefore, the total potential energy is now simplified as

$$V = \sum_i \sum_j q_i K_{ij} q_j = \mathbf{q}^T \mathbf{K} \mathbf{q}, \tag{7.45}$$

which is a quadratic form.

For any small oscillation, the velocity is linear in terms of $\dot{q}_i$, and thus the corresponding kinetic energy is $\frac{1}{2} m \dot{q}_i^2$. The total kinetic energy is the sum of all the components over all particles or parts, forming a quadratic form. That is to say,

$$T = \sum_i \sum_j m_{ij} \dot{q}_i \dot{q}_j = \dot{\mathbf{q}}^T \mathbf{M} \dot{\mathbf{q}}, \tag{7.46}$$

where $\mathbf{M} = [m_{ij}]$ is the mass matrix.

For a conservative system, the total mechanical energy $E = T + V$ is conserved, and thus time-independent. So we have

$$\frac{d(T+V)}{dt} = \frac{d}{dt}[\dot{\mathbf{q}}^T M \dot{\mathbf{q}} + \mathbf{q}^T \mathbf{K} \mathbf{q}] = 0. \tag{7.47}$$

Since $\mathbf{M}$ and $\mathbf{K}$ are symmetric matrices, this above equation becomes

$$\mathbf{M}\ddot{\mathbf{q}} + \mathbf{K}\mathbf{q} = 0. \tag{7.48}$$

This is a second order ordinary differential equation for matrices. Assuming the solution in the form $\mathbf{q}^T = (q_1, q_2, ..., q_n) = (U_1 \cos \omega t, U_2 \cos \omega t, ..., U_n \cos \omega t)$ and substituting it into the above equation, we have

$$|\mathbf{M} - \omega^2 \mathbf{K}| = 0, \tag{7.49}$$

which is an eigenvalue problem.

As an application, let us solve the same system of three mass blocks discussed earlier as shown in Figure 7.6. The total potential energy $T$ is the sum of each mass block

$$T = \frac{1}{2} m_1 (\dot{u}_1)^2 + \frac{1}{2} m_2 (\dot{u}_2)^2 + \frac{1}{2} m_3 (\dot{u}_3)^2$$

$$= \begin{pmatrix} \dot{u}_1 & \dot{u}_2 & \dot{u}_3 \end{pmatrix} \begin{pmatrix} m_1 & 0 & 0 \\ 0 & m_2 & 0 \\ 0 & 0 & m_3 \end{pmatrix} \begin{pmatrix} \dot{u}_1 \\ \dot{u}_2 \\ \dot{u}_3 \end{pmatrix}, \tag{7.50}$$

which can be written as a quadratic form

$$T = \dot{\mathbf{u}}^T \mathbf{M} \dot{\mathbf{u}}, \tag{7.51}$$

where $\mathbf{u}^T = (u_1, u_2, u_3)$, and

$$\mathbf{M} = \frac{1}{2} \begin{pmatrix} m_1 & 0 & 0 \\ 0 & m_2 & 0 \\ 0 & 0 & m_3 \end{pmatrix}. \tag{7.52}$$

We see that $\mathbf{M}$ is a symmetric matrix.

For a spring, the force is $f = kx$, thus the potential energy stored in a spring is

$$\int_0^u kx dx = \frac{1}{2} k u^2. \tag{7.53}$$

Therefore, the total potential energy of the two-spring system is

$$V = \frac{1}{2} k_1 (u_2 - u_1)^2 + \frac{1}{2} k_2 (u_3 - u_2)^2. \tag{7.54}$$

Since interchange of $u_1$ and $u_2$ does not change $V$, it is thus symmetric in terms of $u_1, u_2$ etc, which implies that $K_{ij}$ should be symmetric as well.

The stiffness matrix $\mathbf{K} = [K_{ij}]$ can be calculated using

$$K_{ij} = \frac{1}{2} \frac{\partial^2 V}{\partial u_i \partial u_j}. \tag{7.55}$$

For example,

$$K_{11} = \frac{1}{2} \frac{\partial^2 V}{\partial {u_1}^2} = \frac{1}{2} \times k_1 = \frac{k_1}{2}, \tag{7.56}$$

and

$$K_{12} = \frac{1}{2} \frac{\partial^2 V}{\partial u_1 \partial u_2} = \frac{1}{2} \frac{\partial}{\partial u_1} (\frac{\partial V}{\partial u_2})$$

$$= \frac{1}{2}\frac{\partial}{\partial u_1}[k_1(u_2 - u_1) + k_2(u_3 - u_2)] = -\frac{k_1}{2}. \tag{7.57}$$

Following the similar calculations, we have

$$\mathbf{K} = \frac{1}{2}\begin{pmatrix} k_1 & -k_1 & 0 \\ -k_1 & k_1 + k_2 & k_2 \\ 0 & -k_2 & k_2 \end{pmatrix}, \tag{7.58}$$

which is exactly 1/2 multiplying the stiffness matrix we obtained earlier in equation (7.39). Thus, the equation for small amplitude oscillation is

$$\mathbf{M}\ddot{\mathbf{u}} + \mathbf{K}\mathbf{u} = 0. \tag{7.59}$$

For the special case of $m_1 = m_2 = m_3 = m$ and $k_1 = k_2 = k$, its eigenvalues are determined by

$$\begin{vmatrix} k - \omega^2 m & -k & 0 \\ -k & 2k - \omega^2 m & -k \\ 0 & -k & k - \omega^2 m \end{vmatrix} = 0, \tag{7.60}$$

which is exactly the problem we solved in the previous section (see example 7.1).

For a simple system such as a pendulum, equation (7.39) is equivalent to the following simple formula for calculating the natural frequency

$$\omega = \sqrt{\frac{V''(q)}{M(q)}}, \tag{7.61}$$

where $V'' > 0$ because the potential energy at equilibrium is minimum.

---

□ ***Example 7.2:*** *A simple pendulum with a mass $m$ is hanged vertically from a ceiling with a distance $L$ from the fixed point. Let $\theta$ be the small angle from its equilibrium, then the kinetic energy is $T = \frac{1}{2}mv^2 = \frac{1}{2}mL^2(\dot{\theta})^2$. The potential energy is*

$$V = mgL(1 - \cos\theta).$$

Therefore, the stiffness is $K = \frac{1}{2}V''(\theta) = \frac{1}{2}mgL\cos\theta|_{\theta=0} = mgL/2$. The equivalent mass is $M(\theta) = \frac{1}{2}mL^2$. The governing equation becomes

$$\frac{1}{2}mL^2\ddot{\theta} + \frac{1}{2}mgL\theta,$$

or

$$\ddot{\theta} + \frac{L}{g}\theta = 0.$$

The natural frequency for small oscillations is

$$\omega = \sqrt{\frac{V''}{M(q)}} = \sqrt{\frac{g}{L}}.$$

The period of this pendulum is

$$\tau = \frac{2\pi}{\omega} = 2\pi\sqrt{\frac{L}{g}}.$$

□

# Chapter 8

# Integral Transforms

The mathematical transform is a method of changing one kind of functions and equations into another kind, often simpler or solvable one. In general, the transform is essentially a mathematical operator that produces a new function $F(s)$ by integrating the product of an existing function $f(t)$ and a kernel function $K(t, s)$ between suitable limits

$$F(s) = \int K(t, s) f(t) dt. \tag{8.1}$$

In the Laplace transform, the kernel is simply $\exp(-st)$ and integration limits are from 0 to $\infty$. In the Fourier transform, the kernel is $\exp(\pm ist)$ with a normalized factor $1/\sqrt{2\pi}$.

The Fourier transform maps the time domain of a time-dependent series such as a signal into a frequency domain, which is common practice in signal processing. The Laplace transform is a very powerful tool in solving differential equations. In this chapter, we will focus on the three major transforms: Fourier, Laplace and Wavelet transforms. They are commonly encountered in engineering and computational sciences.

## 8.1 Fourier Transform

### 8.1.1 Fourier Series

For a function $f(t)$ on an interval $t \in [-T, T]$, the Fourier series is defined as

$$f(t) = \frac{a_0}{2} + \sum_{n=1}^{\infty} a_n \cos(\frac{n\pi t}{T}) + b_n \sin(\frac{n\pi t}{T}), \qquad (8.2)$$

where

$$a_0 = \frac{1}{T} \int_{-T}^{T} f(t)dt, \qquad (8.3)$$

and

$$a_n = \frac{1}{T} \int_{-T}^{T} f(t) \cos(\frac{n\pi t}{T})dt, \qquad (8.4)$$

$$b_n = \frac{1}{T} \int_{-T}^{T} f(T) \sin(\frac{n\pi t}{T})dt, \quad (n = 1, 2, ...). \qquad (8.5)$$

Here $a_n$ and $b_n$ are the Fourier coefficients of $f(t)$ on $[-T, T]$. The function $f(t)$ can be continuous or piecewise continuous with a jump discontinuity. For a jump discontinuity at $t = t_0$, if $f'(t_0-)$ and $f'(t_0+)$ both exist, but $f(t_0-) \neq f(t_0+)$, then the Fourier series converges to $[f(t_0-)+f(t_0+)]/2$. The Fourier series in general tends to converge slowly.

From the coefficients $a_n$ and $b_n$, one can easily see that $b_n = 0$ for an even function $f(-t) = f(t)$. Similarly, $a_0 = a_n = 0$ for an odd function $f(-t) = -f(t)$. In both cases, only one-side $[0, T]$ of the integration can be used due to the symmetry. Thus, for an even function $f(t)$, we have the Fourier cosine series on [0,T]

$$f(t) = \frac{a_0}{2} + \sum_{n=1}^{\infty} a_n \cos(\frac{n\pi t}{T}). \qquad (8.6)$$

For an odd function $f(t)$, we have the Fourier sine series

$$f(t) = \sum_{n=1}^{\infty} \sin(\frac{n\pi t}{T}). \qquad (8.7)$$

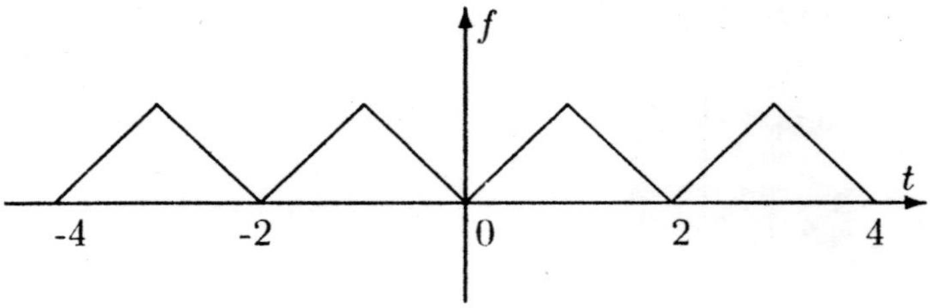


Figure 8.1: Triangular wave with a period of 2.

---

□ ***Example 8.1:*** *The triangular wave is defined by $f(t) = |t|$ for $t \in [-1, 1]$ with a period of 2 or $f(t+2) = f(t)$. Using the coefficients of the Fourier series, we have*

$$a_0 = \frac{1}{2}\int_{-1}^{1} |t|dt = \frac{1}{2}[\int_{-1}^{0}(-t)dt + \int_{0}^{1} tdt] = \frac{1}{2}.$$

*Since both $|t|$ and $\cos(n\pi t)$ are even functions, we have for any $n \geq 1$,*

$$a_n = \int_{-1}^{1} |t|\cos(n\pi t)dt = 2\int_{0}^{1} t\cos(n\pi t)dt$$

$$= 2\frac{t}{n\pi}\sin(n\pi t)\Big|_{0}^{1} - \frac{2}{n\pi}\int_{0}^{1}\sin(n\pi t)dt = \frac{2}{n^2\pi^2}[\cos(n\pi) - 1].$$

*Because $|t|\sin(n\pi t)$ is an odd function, we have*

$$b_n = \int_{-1}^{1} |t|\sin(n\pi t)dt = 0.$$

*Hence, the Fourier series for the triangular wave can be written as*

$$f(t) = \frac{1}{2} + 2\sum_{n=0}^{\infty}\frac{\cos(n\pi) - 1}{n^2\pi^2}\cos(n\pi t).$$

□

---

The $n$-term of the Fourier series, that is $a_n\cos(n\pi t/T) + b_n\sin(n\pi t/T)$, is called the $n$-th harmonic. The energy of the $n$ harmonic is defined by $A_n^2 = a_n^2 + b_n^2$. The sequence of $A_n$ forms the energy or power spectrum of the Fourier series. The energy spectrum of the triangular wave is shown in Figure 8.2.

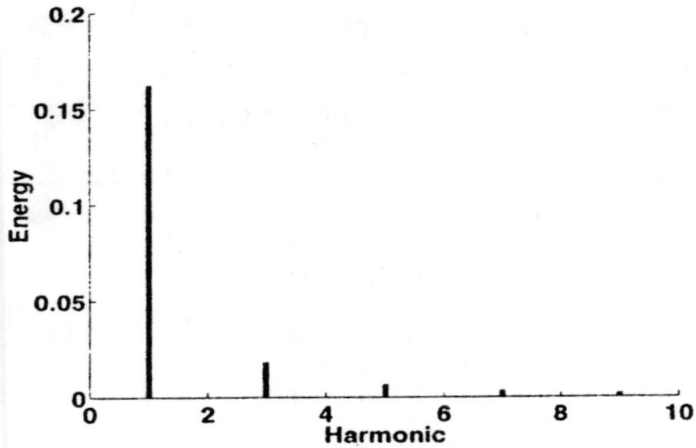


Figure 8.2: Energy spectrum of the triangular wave.

### 8.1.2 Fourier Integral

For the Fourier coefficients of a function defined on the whole real axis $[-\infty, \infty]$, we can take the limits

$$a(\omega_n) = \int_{-T}^{T} f(t)\cos(\omega_n t)dt,$$

and

$$b(\omega_n) = \int_{-T}^{T} f(t)\sin(\omega_n t)dt, \qquad \omega_n = \frac{n\pi}{T}, \tag{8.8}$$

as the limits $T \to \infty$ and $\omega_n \to 0$. We have $a_0 \to 0$ if

$$\int_{\infty}^{\infty} |f(t)| < \infty. \tag{8.9}$$

In this case, the Fourier series becomes the Fourier integral

$$f(t) = \int_0^{\infty} [a(\omega)\cos(\omega t) + b(\omega)\sin(\omega t)d\omega, \tag{8.10}$$

where

$$a(\omega) = \frac{1}{\pi}\int_{-\infty}^{\infty} f(t)\cos(\omega t)dt, \tag{8.11}$$

$$b(\omega) = \frac{1}{\pi}\int_{-\infty}^{\infty} f(t)\sin(\omega t)dt. \tag{8.12}$$

Following the similar discussion for even and odd functions, we know that even functions lead to Fourier cosine integrals and odd functions lead to Fourier sine integrals.

### 8.1.3 Fourier Transform

The Fourier transform $\mathcal{F}[f(t)]$ of $f(t)$ is defined as

$$F(\omega) = \mathcal{F}[f(t)] = \frac{1}{\sqrt{2\pi}} \int_{-\infty}^{\infty} f(t) e^{-i\omega t} dt, \tag{8.13}$$

and the inverse Fourier transform can be written as

$$f(t) = \mathcal{F}^{-1}[F(\omega)] = \frac{1}{\sqrt{2\pi}} \int_{-\infty}^{\infty} F(\omega) e^{i\omega t} d\omega, \tag{8.14}$$

where $\exp[i\omega t] = \cos(\omega t) + i \sin(\omega t)$. The Fourier transform is a linear operator, and it has most of the properties of the differential operator and the integral operator. Hence, it is straightforward to prove that it has the following properties:

$$\mathcal{F}[f(t) + g(t)] = \mathcal{F}[f(t)] + \mathcal{F}[g(t)], \tag{8.15}$$

$$\mathcal{F}[\alpha f(t)] = \alpha \mathcal{F}[f(t)], \tag{8.16}$$

$$\mathcal{F}[(-it)^n f(t)] = \frac{d^n F(\omega)}{d\omega^n}, \tag{8.17}$$

and

$$\mathcal{F}[f^{(n)}(t)] = (i\omega)^n F(\omega), \tag{8.18}$$

if $f(t \to \pm\infty) = f'(t \to \pm\infty) = ... = f^{(n-1)}(t \to \pm\infty) \to 0$. The transform can have different variations such as the Fourier sine transform and the Fourier cosine transform. The Fourier transforms of some common functions are listed in the following table 8.1.

---

□ ***Example 8.2:*** *For the triangle function $f(t) = 1 - |t|$ for $(|t| \leq 1)$ and $f(t) = 0$ for $|t| > 1$, its Fourier transform is*

$$\sqrt{2\pi} F(\omega) = \int_{-\infty}^{\infty} (1 - |t|) e^{-i\omega t} dt$$

$$= \int_{-1}^{0} (1+t)e^{-i\omega t}dt + \int_{0}^{1} (1-t)e^{-i\omega t}dt.$$

*Integrating by parts and using*

$$\cos\omega = (e^{i\omega} + e^{-i\omega})/2, \qquad \sin^2(\omega/2) = \frac{1-\cos\omega}{2},$$

*we have*

$$\sqrt{2\pi}F(\omega) = \frac{i\omega + 1 - e^{i\omega}}{\omega^2} - \frac{e^{-i\omega} + i\omega - 1}{\omega^2} = \frac{2(1-\cos\omega)}{\omega^2}.$$

*Hence, we have*

$$F(\omega) = \frac{1}{\sqrt{2\pi}} \frac{\sin^2(\omega/2)}{(\omega/2)^2} = \frac{1}{\sqrt{2\pi}} \mathrm{sinc}^2(\omega/2).$$

□

Table 8.1: Fourier Transforms

| f(t) | $F(\omega) = \mathcal{F}[f(t)]$ |
|---|---|
| $f(t-t_0)$ | $F(\omega)e^{-i\omega t_0}$ |
| $f(t)e^{-i\omega_0 t}$ | $F(\omega - \omega_0)$ |
| $\delta(t)$ | $1/\sqrt{2\pi}$ |
| 1 | $\sqrt{2\pi}\delta(\omega)$ |
| $\mathrm{sign}(t)$ | $\frac{2}{i\omega}$ |
| $e^{-\alpha\|t\|}$ | $\frac{2\alpha}{\alpha^2+\omega^2}$ |
| $e^{-(\alpha t)^2}$ $(\alpha > 0)$ | $\frac{1}{\sqrt{2}\alpha} e^{-\frac{\omega^2}{4\alpha^2}}$ |
| $f(\alpha t)$ | $\frac{1}{\|\alpha\|}F(\frac{\omega}{\alpha})$ |
| $\frac{1}{\alpha^2+t^2}$ | $\sqrt{\frac{\pi}{2}} \frac{e^{-\alpha\|\omega\|}}{\alpha}$ |
| $\cos(\omega_0 t)$ | $\sqrt{\frac{\pi}{2}}[\delta(\omega-\omega_0) + \delta(\omega+\omega_0)]$ |
| $\sin(\omega_0)$ | $i\sqrt{\frac{\pi}{2}}[\delta(\omega+\omega_0) - \delta(\omega-\omega_0)]$ |
| $\frac{\sin\alpha x}{x}$ $(\alpha > 0)$ | $\sqrt{\frac{\pi}{2}}$, $(\|\omega\| < \alpha); 0$, $(\|\omega\| > \alpha)$ |

The most useful Fourier transform for engineering and computational science is probably the discrete form, especially in digital signal processing. The discrete Fourier transform (DFT), for a periodic discrete function or signal $x(n)$ with a period $N$, is defined by

$$X[k] = \sum_{n=0}^{N-1} x[n]e^{-i\frac{2\pi kn}{N}}, \tag{8.19}$$

and the inverse transform, also called the signal reconstruction, is defined by

$$x[n] = \frac{1}{N}\sum_{k=0}^{N-1} X[k]e^{i\frac{2\pi kn}{N}}. \tag{8.20}$$

A periodic signal $x(n+N) = x(n)$ has a periodic spectrum $X[k+N] = X[k]$. The discrete Fourier transform consists of $N$ multiplications and $N-1$ additions for each $X[k]$, thus for $N$ values of $k$, the computational complexity is of $O(N^2)$. However, if $N = 2^m$ ($m \in \mathcal{N}$), many of the DFT calculations are not necessary. In fact, by rearranging the formula, one can get the complexity of $O(N \log_2 N)$. This type of algorithms is called Fast Fourier Transform (FFT). Vast literature exists on the signal processing such as FFT, filter design and signal reconstruction.

## 8.2 Laplace Transforms

The Laplace transform $\mathcal{L}[f(t)]$ of a function $f(t)$ is defined as

$$F(s) = \mathcal{L}[f(t)] = \int_0^{\infty} f(t)e^{-st}dt, \tag{8.21}$$

where $s > 0$. The inverse Laplace transform $\mathcal{L}^{-1}[F(s)]$ is $f(t)$ or $f(t) = \mathcal{L}^{-1}[F(s)]$. The Laplace transforms of most simple functions can be obtained by direct integration. For simple functions $t$ and $e^{\alpha t}$, we have

$$\mathcal{L}[t] = \int_0^{\infty} te^{-st}dt = \int_0^{\infty} \frac{1}{s}e^{-st}dt + \left[-\frac{t}{s}e^{-st}\right]_0^{\infty} = \frac{1}{s^2}.$$

$$\mathcal{L}[e^{\alpha t}] = \int_0^{\infty} e^{\alpha t}e^{-st}dt = \left[-\frac{1}{s-\alpha}e^{-(s-\alpha)t}\right]_0^{\infty} = \frac{1}{s-\alpha}.$$

Conversely, $\mathcal{L}^{-1}[\frac{1}{s^2}] = t$, $\quad \mathcal{L}^{-1}[\frac{1}{s-\alpha}] = e^{\alpha t}$. For the Dirac $\delta$-function, we have its Laplace transform

$$\mathcal{L}[\delta(t)] = \int_0^\infty \delta(t) e^{-st} dt = e^{-st}\Big|_{t=0} = 1. \qquad (8.22)$$

However, the inverse of a Laplace transform is usually more complicated. It often involves the partial fractions of polynomials and usage of different rules of Laplace transforms. From the basic definition, it is straightforward to prove that the Laplace transform has the following properties:

$$\mathcal{L}[\alpha f(t) + \beta g(t)] = \alpha \mathcal{L}[f(t)] + \beta \mathcal{L}[g(t)], \qquad (8.23)$$

$$\mathcal{L}[e^{\alpha t} f(t)] = F(s-\alpha), \quad s > \alpha, \qquad (8.24)$$

$$\mathcal{L}[f(t-\alpha)] = e^{-\alpha s} \mathcal{L}[f(t)], \qquad (8.25)$$

$$\mathcal{L}[f'(t)] = s\mathcal{L}[f(t)] - f(0), \qquad (8.26)$$

$$\mathcal{L}[\int_0^t f(\tau) d\tau] = \frac{1}{s} \mathcal{L}[f], \qquad (8.27)$$

The Laplace transform pairs of common functions are listed below in table 8.2.

---

□ ***Example 8.3:*** *In order to obtain the Laplace transform of* $f(t) = \cos \omega t$, *we shall first write*

$$f(t) = \cos \omega t = \frac{1}{2}(e^{i\omega t} + e^{-i\omega t}).$$

*Then, we have*

$$\mathcal{L}[f(t)] = F(s) = \int_0^\infty [\frac{1}{2}(e^{i\omega t} + e^{-i\omega t})] e^{-st} dt$$

$$= \frac{1}{2}[\int_0^\infty e^{(-s+i\omega)t} dt + \int_0^\infty e^{-s-i\omega)t} dt]$$

$$= \frac{1}{2}[\frac{1}{s-i\omega} + \frac{1}{s+i\omega}] = \frac{s}{s^2+\omega^2}.$$

□

---

Table 8.2: Laplace Transform

| Function $f(t)$ | Laplace Transform $F(s)$ |
|---|---|
| $1$ | $\frac{1}{s}$ |
| $\delta(t)$ | $1$ |
| $t^n, n > 0$ | $\frac{n!}{s^{n+1}}$ |
| $\cos(\alpha t)$ | $\frac{s}{s^2+\alpha^2}$ |
| $\sin(\alpha t)$ | $\frac{\alpha}{s^2+\alpha^2}$ |
| $e^{\alpha t}$ | $\frac{1}{s-\alpha}$ |
| $t^{1/2}$ | $\frac{1}{2}(\frac{\pi}{s^3})^{1/2}$ |
| $t^{-1/2}$ | $\sqrt{\frac{\pi}{s}}$ |
| $t^n f(t)$ | $(-1)^n \frac{d^n F(s)}{ds^n}$ |
| $\cos(\alpha t + \beta)$ | $\frac{s\cos(\beta)-\alpha\sin(\beta)}{s^2+\alpha^2}$ |
| $\sinh(\alpha t)$ | $\frac{\alpha}{s^2-\alpha^2}$ |
| $\cosh(\alpha t)$ | $\frac{s}{s^2-\alpha^2}$ |
| $\mathrm{erfc}(\frac{\alpha}{2\sqrt{t}})$ | $\frac{1}{s}e^{-\alpha\sqrt{s}}$ |
| $\frac{1}{\sqrt{\pi t}}e^{-\frac{\alpha^2}{4t}}$ | $\frac{1}{\sqrt{s}}e^{-\alpha\sqrt{s}}$ |
| $\sin\alpha\sqrt{t}$ | $\frac{\alpha}{2}\sqrt{\frac{\pi}{s^3}}e^{-\frac{\alpha^2}{4s}}$ |
| $\frac{1-e^{-\alpha t}}{t}$ $(\alpha > 0)$ | $\ln(1+\frac{\alpha}{s})$ |
| $\frac{1}{\alpha-\beta}(e^{\alpha t}-e^{\beta t})$ $(\alpha \neq \beta)$ | $\frac{1}{(s-\alpha)(s-\beta)}$ |

Both Fourier and Laplace transforms follow the convolution theorem. For two functions $f$ and $g$, their convolution $f * g$ obeys

$$f * g = \int_0^t f(t-\alpha)g(\alpha)d\alpha. \tag{8.28}$$

and their Laplace transforms follow

$$\mathcal{L}[f(t) * g(t)] = F(s)G(s), \tag{8.29}$$

$$\mathcal{L}^{-1}[F(s)G(s)] = \int_0^t f(t-\alpha)g(\alpha)d\alpha. \tag{8.30}$$

The Fourier transform has the similar properties

$$f(t) * g(t) = \int_{-\infty}^{\infty} f(t)g(t-u)du. \tag{8.31}$$

$$\mathcal{F}[f(t) * g(t)] = F(\omega)G(\omega). \tag{8.32}$$

## 8.3 Wavelet

The Fourier transform is an ideal tool for studying the stationary time signal whose properties are statistically invariant over time. In the Fourier transform, the stationary signal is decomposed into linear combinations of sine and cosine waves

$$\frac{1}{\sqrt{2\pi}}, \quad \frac{1}{\sqrt{\pi}}\cos(nt), \quad \frac{1}{\sqrt{\pi}}\sin(nt), \quad (n = 1, 2, ...). \tag{8.33}$$

The Fourier transform is very useful to analyse stationary signals where a stationary signal means that the frequencies of the signal do not change with time. For non-stationary signals whose frequencies $f = \omega/2\pi$ vary with time (see Figure 8.3), the Fourier transform does not work well. In addition, in the Fourier transform there is a tradeoff between frequency resolution and time resolution,

$$\Delta w \Delta t \geq \frac{1}{2}, \tag{8.34}$$

which is similar to the Heisenberg uncertainty principle for spatial and velocity intervals. The wavelet transform is an alternative approach to the Fourier transform to overcome the resolution problem using the Mother wavelet $\psi$ or prototype for generating the other windows functions, and all the used windows are in the form of either dilated/compressed or shifted. As a result, the wavelet transform is very powerful in dealing with non-stationary signals because the Fourier transform is not suitable for such signals. In the wavelet transform, a transient signal is decomposed into elementary components of wavelets

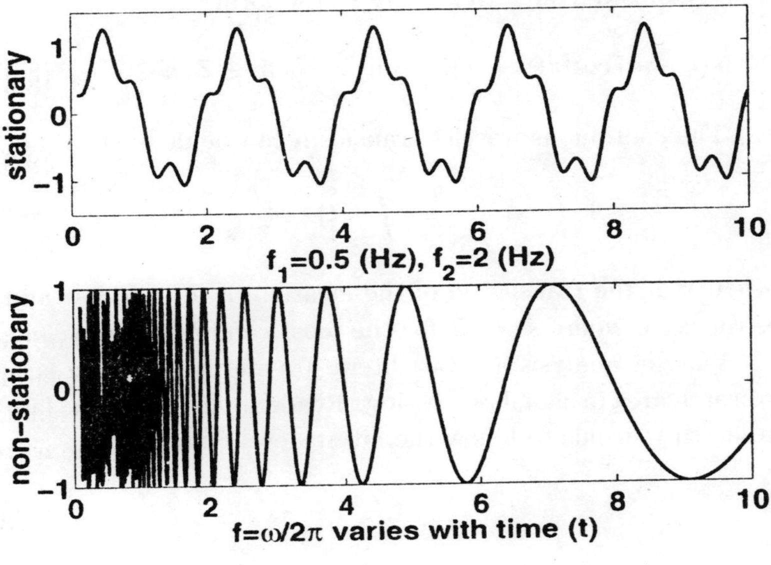


Figure 8.3: A stationary signal with two fixed frequencies ($f_1 = 0.5$ Hz and $f_2 = 2$ Hz) and a non-stationary signal whose frequency varies with time.

or wavelet packets. There are three major type of wavelets: Grossmann-Morlet wavelets, Daubechies wavelets and Gabor-Malvar wavelets.

Wavelets are defined as a real-valued function $\psi(t)$ ($t \in \mathcal{R}$) in terms of the generator wavelet or mother wavelet. The function $\psi$ is both well localized, decreasing rapidly as $t \to \infty$ and oscillating in a wavery manner. To generate other wavelets, $\psi(\alpha, \beta, t)$ is used by translating in time and change of scales.

Grossmann-Morlet wavelets are of the form

$$\frac{1}{\alpha}\psi(\frac{t-\beta}{\alpha}), \quad \alpha > 0, \qquad a, b \in \mathcal{R}, \tag{8.35}$$

where $\psi$ a generator wavelet. The Daubechies wavelets have the form

$$2^{n/2}\psi(2^n t - m), \qquad m, n \in \mathcal{Z}. \tag{8.36}$$

The Gabor-Malvar wavelets are in the form

$$w(t-m)\cos[\pi(n+\frac{1}{2})(t-m)], \qquad m \in \mathcal{Z}, n \in N. \tag{8.37}$$

The continuous wavelet transform can be defined by

$$\Psi_f(\tau, s) = \frac{1}{\sqrt{|s|}} \int f(t) \cdot \psi(\frac{t-\tau}{s})dt, \tag{8.38}$$

where $\tau$ is the translation of the location of the window and $s$ is the scale where $s = 1$ is for the most compressed wavelet.

Wavelet analysis has vast literature and it is still an active research area in signal processing. Readers can search the latest research journals to follow the latest developments.

# Chapter 9

# Partial Differential Equations

Partial differential equations are much more complicated compared with the ordinary differential equations. There is no universal solution technique for nonlinear equations, even the numerical simulations are usually not straightforward. Thus, we will mainly focus on the linear partial differential equations and the equations of special interests in engineering and computational sciences. A partial differential equation (PDE) is a relationship containing one or more partial derivatives. Similar to the ordinary differential equation, the highest $n$th partial derivative is referred to as the order $n$ of the partial differential equation. The general form of a partial differential equation can be written as

$$\psi(x, y, ..., \frac{\partial u}{\partial x}, \frac{\partial u}{\partial y}, \frac{\partial^2 u}{\partial x^2}, \frac{\partial^2 u}{\partial y^2}, \frac{\partial^2 u}{\partial x \partial y}, ...) = 0. \tag{9.1}$$

where $u$ is the dependent variable and $x, y, ...$ are the independent variables.

A simple example of partial differential equations is the linear first order partial differential equation, which can be written as

$$a(x, y)\frac{\partial u}{\partial x} + b(x, y)\frac{\partial u}{\partial y} = f(x, y). \tag{9.2}$$

for two independent variables and one dependent variable $u$. If the right hand side is zero or simply $f(x, y) = 0$, then the equation is said to be homogeneous. The equation is said to be linear if $a, b$ and $f$ are functions of $x, y$ only, not $u$ itself.

For simplicity in notations in the studies of partial differential equations, compact subscript forms are often used in the literature. They are

$$u_x \equiv \frac{\partial u}{\partial x}, \quad u_y \equiv \frac{\partial u}{\partial y}, \quad u_{xx} \equiv \frac{\partial^2 u}{\partial x^2},$$

$$u_{yy} \equiv \frac{\partial^2 u}{\partial y^2}, \quad u_{xy} \equiv \frac{\partial^2 u}{\partial x \partial y}, \quad \ldots \tag{9.3}$$

and thus we can write (9.2) as

$$au_x + bu_y = f. \tag{9.4}$$

In the rest of the chapters in this book, we will use these notations whenever no confusion occurs.

## 9.1 First Order PDE

The first order partial differential equation of linear type can be written as

$$a(x, y)u_x + b(x, y)u_y = f(x, y), \tag{9.5}$$

which can be solved using the method of characteristics

$$\frac{dx}{a} = \frac{dy}{b} = \frac{du}{f}. \tag{9.6}$$

This is equivalent to the following equation in terms of parameter $s$

$$\frac{dx}{ds} = a, \quad \frac{dy}{ds} = b, \quad \frac{du}{ds} = f, \tag{9.7}$$

which essentially forms a system of first-order ordinary differential equations.

The simplest example of a first order linear partial differential equation is the first order hyperbolic equation

$$u_t + cu_x = 0, \tag{9.8}$$

where $c$ is a constant. It has a general solution of

$$u = \psi(x - ct), \tag{9.9}$$

which is a travelling wave along $x$-axis with a constant speed $c$. If the initial shape is $u(x,0) = \psi(x)$, then $u(x,t) = \psi(x - ct)$ at time $t$, therefore the shape of the wave does not change though its position is constantly changing.

## 9.2 Classification

A linear second-order partial differential equation can be written in the generic form in terms of two independent variables $x$ and $y$,

$$au_{xx} + bu_{xy} + cu_{yy} + gu_x + hu_y + ku = f, \tag{9.10}$$

where $a, b, c, g, h, k$ and $f$ are functions of $x$ and $y$ only. If $f(x, y, u)$ is also a function of $u$, then we say that this equation is quasi-linear.

If $\Delta = b^2 - 4ac < 0$, the equation is elliptic. One famous example is the Laplace equation $u_{xx} + u_{yy} = 0$.

If $\Delta > 0$, it is hyperbolic. One example is the wave equation $u_{tt} = c^2 u_{xx}$.

If $\Delta = 0$, it is parabolic. Diffusion and heat conduction equations are of the parabolic type $u_t = \kappa u_{xx}$.

## 9.3 Classic PDEs

Many physical processes in engineering are governed by three classic partial differential equations so they are widely used in a vast range of applications.

### Laplace's and Poisson's Equation

In heat transfer problems, the steady state of heat conduction with a source is governed by the Poison equation

$$k\nabla^2 u = f(x, y, t), \qquad (x, y) \in \Omega, \tag{9.11}$$

or

$$u_{xx} + u_{yy} = q(x, y, t), \tag{9.12}$$

for two independent variables $x$ and $y$. Here $k$ is the thermal diffusivity and $f(x, y, t)$ is the heat source. If there is no heat source ($q = 0$), this becomes the Laplace equation. The solution or a function is said to be harmonic if it satisfies the Laplace equation.

### Heat Conduction Equation

Time-dependent problems, such as diffusion and transient heat conduction, are governed by parabolic equations. The heat conduction equation

$$u_t = k u_{xx}, \tag{9.13}$$

is a famous example. For diffusion problem, $k$ is replaced by the diffusion coefficient $D$.

### Wave Equation

The vibrations of strings and travelling sound waves are governed by the hyperbolic wave equation. The 1-D wave equation in its simplest form is

$$u_{tt} = c^2 u_{xx}, \tag{9.14}$$

where $c$ is the speed of the wave.

There are other equations that occur frequently in mathematical physics, engineering and computational sciences. We will give a brief description of some of these equations in later chapters.

# Chapter 10

# Techniques for Solving PDEs

Different types of equations usually require different solution techniques. However, there are some methods that work for most of the linearly partial differential equations with appropriate boundary conditions on a regular domain. These methods include separation of variables, series expansions, similarity solutions, hybrid methods, and integral transform methods.

## 10.1 Separation of Variables

The separation of variables attempts a solution of the form

$$u = X(x)Y(y)T(t), \tag{10.1}$$

where $X(x), Y(y), T(t)$ are functions of $x, y, t$, respectively. In order to determine these functions, they have to satisfy the partial differential equation and the required boundary conditions. As a result, the partial differential equation is usually transformed into two or three ordinary differential equations (ODEs), and these ordinary differential equations often appear as eigenvalue problems. The final solution is then obtained by solving these ODEs. As a classic example, we now try to solve the 1-D heat conduction equation in the domain $x \in [0, L]$ and $t > 0$

$$u_t = k u_{xx}. \tag{10.2}$$

with the initial value and boundary conditions

$$u(0, t) = u(L, t) = 0, \qquad u(x, 0) = \psi. \tag{10.3}$$

Letting $u(x, t) = X(x)T(t)$, we have

$$\frac{X''(x)}{X} = \frac{T'(t)}{kT}. \tag{10.4}$$

As the left hand side depends only on $x$ and the right hand side only depends on $t$, therefore, both sides must be equal to the same constant, and the constant can taken to be as $-\lambda^2$. The negative sign is just for convenience because we will see below that the finiteness of the solution $T(t)$ requires that eigenvalues $\lambda^2 > 0$ or $\lambda$ are real. Hence, we now get two ordinary differential equations

$$X''(x) + \lambda^2 X(x) = 0, \qquad T'(t) + k\lambda^2 T(t) = 0, \tag{10.5}$$

where $\lambda$ is the eigenvalue. The solution for $T(t)$ is

$$T = A_n e^{-k\lambda^2 t}. \tag{10.6}$$

The solution for $X(x)$ is in a generic form

$$X(t) = \alpha \cos \lambda x + \beta \sin \lambda x. \tag{10.7}$$

From the boundary condition $u(0, t) = 0$, we have $\alpha = 0$. From $u(L, t) = 0$, we have

$$\sin \lambda L = 0, \tag{10.8}$$

which requires that $\lambda L = n\pi$. Please note that $n \neq 0$ because the solution is trivial if $n = 0$. Therefore, $\lambda$ cannot be continuous, and it only takes an infinite number of discrete values, called eigenvalues. Each eigenvalue $\lambda = \lambda_n = n\pi/L$, $(n =$

1, 2, ...) has a corresponding eigenfunction $X_n = \sin(\lambda_n x)$. Substituting into the solution for $T(t)$, we have

$$T(t) = A_n e^{-\lambda^2 kt} = A_n e^{-\frac{(n\pi)^2}{L^2} kt} \tag{10.9}$$

By superimposing $u_n = X_n T_n$ and expanding the initial condition into a Fourier series so as to determine the coefficients, we have

$$u(x, t) = \sum_{n=1}^{\infty} \alpha_n \sin(\frac{n\pi x}{L}) e^{-(\frac{n\pi}{L})^2 kt},$$

$$\alpha_n = \frac{2}{L} \int_0^L \psi(x) \sin(\frac{n\pi d\xi}{L}) d\xi. \tag{10.10}$$

## 10.2 Transform Methods

### Laplace Transform

The basic idea of the integral transform method is to reduce the number of the independent variables. For the 1-D time-dependent heat conduction, it transforms the partial differential equation into an ordinary differential equation. By solving the ordinary differential equation, the solution to the original problem is obtained by inverting back from the Laplace transform. As an example, we now solve the 1-D heat conduction in semi-infinite interval $[0, \infty)$,

$$\frac{\partial T}{\partial t} = k \frac{\partial^2 T}{\partial x^2}, \tag{10.11}$$

with the boundary conditions

$$T(x, 0) = 0, \quad T(0, t) = T_0. \tag{10.12}$$

Let $\bar{T}(x, s) = \int_0^\infty T(x, t) e^{-st} dt$ be the Laplace transform of $T(x, t)$, the equation then becomes

$$s\bar{T} = k \frac{d^2 \bar{T}}{dx^2}, \tag{10.13}$$

and the boundary condition at $x = 0$ becomes $\bar{T}_{x=0} = T_0/s$. The general solution to the ordinary differential equation can be written as

$$\bar{T} = Ae^{-\sqrt{\frac{s}{k}}x} + Be^{\sqrt{\frac{s}{k}}x}.$$

The finiteness of the solution as $x \to \infty$ requires that $B = 0$, and the boundary conditions lead to

$$\bar{T} = \frac{T_0}{s} e^{-\sqrt{\frac{s}{k}}x}.$$

By the inversion of the Laplace transform, we have

$$T = T_0 \text{erfc}(\frac{x}{2\sqrt{kt}}),$$

where erfc$(x)$ is the complementary error function.

## Fourier Transform

Fourier transform works in the similar manner as the Laplace transform. The famous example is the classical wave equation

$$u_{tt} = v^2 u_{xx}, \tag{10.14}$$

with the initial conditions $u(x, 0) = \psi(x) = \exp[-(x - a)^2]$, and $u_t(x, 0) = 0$. Let $\bar{u}(\omega, t) = \frac{1}{\sqrt{2\pi}} \int_{-\infty}^{\infty} u(x, t) e^{i\omega x} dx$ be the Fourier transform of $u(x, t)$. This transforms the PDE problem into an ODE

$$\frac{d^2 \bar{u}}{dt^2} = -v^2 \omega^2 \bar{u}, \tag{10.15}$$

with

$$\bar{u} = \bar{\psi}(\omega), \quad \frac{d\bar{u}(\omega, 0)}{dt} = 0. \tag{10.16}$$

The general solution in terms of the parameter $\omega$ is

$$\bar{u}(\omega, t) = \bar{\psi}(\omega) \cos(v\omega t).$$

By using the inverse Fourier transform, we finally have

$$u(x, t) = \frac{1}{\sqrt{2\pi}} \int_{-\infty}^{\infty} \bar{\psi}(\omega) \cos(v\omega t) e^{-i\omega x} d\omega$$

$$= \frac{1}{2}[\psi(x+vt)+\psi(x-vt)]$$

$$= \frac{1}{2}[e^{-(x-a+vt)^2}+e^{-(x-a-vt)^2}], \tag{10.17}$$

which implies two travelling waves: one travels along the $x$-axis and the other along the negative $x$-axis direction.

## 10.3 Similarity Solution

The essence of similarity solution is to use the so-called similarity variable $\xi = x/t^\beta$ so as to reduce the partial differential equation to an ordinary differential equation. For example, the diffusion equation

$$u_t = \kappa u_{xx}, \tag{10.18}$$

can be solved by using the similarity method by defining a similar variable

$$\eta = \frac{x^2}{\kappa t} \tag{10.19}$$

or

$$\zeta = \frac{x}{\sqrt{\kappa t}}. \tag{10.20}$$

In general, we can assume that the solution to the equation has the form

$$u = (\kappa t)^\alpha f[\frac{x}{(\kappa t)^\beta}]. \tag{10.21}$$

By substituting it into the diffusion equation, the coefficients $\alpha$ and $\beta$ can be determined. For most applications, one can assume $\alpha = 0$ so that $u = f(\zeta)$ (see the following example for details). In this case, we have

$$-\zeta\beta(\kappa t)^{-1}f' = \frac{f''}{(\kappa t)^{2\beta}}, \tag{10.22}$$

or

$$f'' = -\zeta f'\beta(\kappa t)^{2\beta-1}, \tag{10.23}$$

where $f' = df/d\zeta$. In deriving this equation, we have used the chain rules of differentiations $\frac{\partial}{\partial x} = \frac{\partial}{\partial \zeta}\frac{\partial \zeta}{\partial x}$ and $\frac{\partial}{\partial t} = \frac{\partial}{\partial \zeta}\frac{\partial \zeta}{\partial t}$ so that

$$\frac{\partial u}{\partial t} = -\frac{\beta k f'(\zeta) x}{(\kappa t)^{\beta} t} = -\beta \zeta f'(\zeta)(\kappa t)^{-1}, \tag{10.24}$$

and

$$\frac{\partial^2 u}{\partial x^2} = \frac{f''(\zeta)}{(\kappa t)^{2\beta}}. \tag{10.25}$$

Since the original equation does not have time-dependent terms *explicitly*, this means that all the exponents for any $t$-terms must be zero. Therefore, we have $2\beta = 1$, or $\beta = \frac{1}{2}$. Now, the diffusion equation becomes

$$f''(\zeta) = -\frac{\zeta}{2} f', \tag{10.26}$$

Using $(\ln f')' = f''/f'$ and integrating the above equation once, we get

$$\ln f' = -\frac{\zeta^2}{4}, \qquad \text{or} \qquad f' = K e^{-\zeta^2/4}. \tag{10.27}$$

Integrating it again and using the $\zeta^2 = 4\xi^2$, we obtain

$$u = A \int_{\xi_0}^{\xi} e^{-\xi^2} d\xi = C\mathrm{erf}(\frac{x}{\sqrt{4\kappa t}}) + D, \tag{10.28}$$

where $C$ and $D$ are constants that can be determined from appropriate boundary conditions. For the same problem as (10.12), the boundary condition as $x \to \infty$ implies that $C+D = 0$, while $u(0,t) = T_0$ means that $D = -C = T_0$. Therefore, we finally have

$$u = T_0[1 - \mathrm{erf}(\frac{x}{\sqrt{4\kappa t}})] = T_0\mathrm{erfc}(\frac{x}{\sqrt{4\kappa t}}). \tag{10.29}$$

---

□ ***Example 10.1:*** *For the similarity solution of the diffusion equation $u_t = \kappa u_{xx}$, we assume $u = (\kappa t)^{\alpha} f(\frac{x}{\kappa t)^{\beta}}$. Now we want to know why $\alpha = 0$. Since*

$$u_t = \alpha\kappa(\kappa t)^{\alpha-1} f - (\kappa t)^{\alpha-1}\kappa\beta\zeta f', \qquad u_{xx} = (\kappa t)^{\alpha}\frac{f''}{(\kappa t)^{2\beta}},$$

*the diffusion equation becomes*

$$(\kappa t)^{\alpha}[\alpha f - \beta \zeta f'] - (\kappa t)^{\alpha - 2\beta + 1} f'' = 0.$$

*The requirement that no explicit (time $t$) appears in the equation leads to*

$$\alpha = 0, \qquad \alpha - 2\beta + 1 = 0.$$

*Thus,*

$$\alpha = 0, \qquad \beta = \frac{1}{2}.$$

□

You may think that why not divide both sides of the above equation by $(\kappa t)^{\alpha}$, then we do not impose any requirement on $\alpha$, thus $\beta = \frac{1+\alpha}{2}$. This non-zero $\alpha$ indeed appears in some nonlinear diffusions equations where the diffusion coefficient is not a constant and $\kappa(u)$ may depend on $u$ itself.

## 10.4 Travelling Wave Solution

The travelling wave technique can be demonstrated using the famous Korteweg-de Vries (KdV) equation

$$\frac{\partial u}{\partial t} + 6u\frac{\partial u}{\partial x} + \frac{\partial^3 u}{\partial x^3} = 0, \tag{10.30}$$

or

$$u_t + 3(u^2)_x + u_{xxx} = 0, \tag{10.31}$$

which is a third-order nonlinear partial differential equation. The interesting feature of this equation is that it has a solitary wave solution or soliton. The soliton phenomenon was first observed by John Russell in 1834 when he travelled along the Union canal in Scotland. Nowadays, telecommunications use solitons to carry signals in optical fibres.

Now we seek the travelling wave solution in the form

$$u = \phi(x - vt). \tag{10.32}$$

where $v$ is the speed of the travelling wave. By substituting into the KdV equation, we have

$$\phi''' + 6\phi\phi' - v\phi = 0. \tag{10.33}$$

Using $(\phi^2/2)' = \phi\phi'$ and integrating the above equation once, we have

$$\phi'' + 3\phi^2 - v\phi = A, \tag{10.34}$$

where $A$ is an integration constant. The requirement of $\phi$, $\phi'$, $\phi'' \to 0$ at far field $x \to \pm\infty$ leads to $A = 0$. Let $\psi = \phi'$, we get

$$\psi \frac{d\psi}{d\phi} + 3\phi^2 - v\phi = 0. \tag{10.35}$$

Integrating with respect to $\phi$, we get

$$\frac{1}{2}\psi^2 = -\phi^3 + \frac{1}{2}v\phi^2. \tag{10.36}$$

Integrating it again and substituting back to $u$, we have the travelling wave solution

$$u = \frac{v}{2}\text{sech}^2[\frac{\sqrt{v}}{2}(x - vt - \delta)], \tag{10.37}$$

where $\delta$ is a constant and $v/2$ is the amplitude of the wave. We can see that the speed of the wave depends on the amplitude or height of the wave. That is to say, big waves travel faster than smaller waves. For linear wave equations, waves can travel in both directions, but here it is only possible for the soliton to travel in one direction, that is along $x$-axis direction in this scenario.

## 10.5 Green's Function

The method of Green's function is very powerful in solving elliptic equations. A Green's function inside the domain $\Omega$ is defined as

$$\nabla^2 G = \Delta G = 2\pi\delta(x - \xi)\delta(y - \eta)\delta(z - \delta), \tag{10.38}$$

where $\delta(x)$ is the Dirac delta function. It usually requires that $G = 0$ on the boundary surface $\Gamma$. Generally speaking, $G = G(x, y, z; \xi, \eta, \zeta)$.

For the hyperbolic equations such as the wave equation, we can define the Green's function as

$$G_{tt} - c^2 \Delta G = \delta(x - \xi)\delta(y - \eta)\delta(z - \zeta)\delta(t - \tau). \qquad (10.39)$$

The fundamental Green's function for this case is

$$G(x, y, z, t; \xi, \eta, \zeta, \tau) = \frac{1}{4\pi c^2 r}\delta[(t - \tau) - \frac{r}{c}]. \qquad (10.40)$$

The Green's function method is very complicated, but it can be very neat in obtaining solutions. Readers can refer to the literature listed at the end of the book.

## 10.6 Hybrid Method

Some differential equations can be solved using one of the methods described above, but often a single method simply does not work. In this case, a hybrid method that combines several methods is needed. For example, the Crank's diffusion problem in an infinite cylinder is governed by the following equation:

$$\frac{\partial u}{\partial t} = \frac{1}{r}\frac{\partial}{\partial r}[r\kappa \frac{\partial u}{\partial r}], \qquad (10.41)$$

where $\kappa$ is the diffusion coefficient, and $r$ is the distance in the polar coordinates. $u(r, t)$ can be concentration or any other quantity. Now we want to solve this equation with the following boundary conditions:

$$u(r, t) = 0, \qquad r = a, \qquad (10.42)$$

and

$$u(r, 0) = \psi(r), \qquad r \in (0, a). \qquad (10.43)$$

First, we use the separation of variables, we have

$$u(r, t) = v(r)e^{-\lambda^2 \kappa t} \qquad (10.44)$$

then we have the equation for $v(r)$

$$v'' + \frac{1}{r}v' + \lambda^2 v = 0, \tag{10.45}$$

where $v' = dv/dr$ and $\lambda$ is a parameter to be determined. This equation is essentially the Bessel equation of the zero-order $\nu = 0$. Thus, we can now assume that the general solution for $u(r,t)$ has the following form

$$u(r,t) = \sum_{i=1}^{\infty} D_i J_0(\lambda_i r) e^{-\lambda_i^2 \kappa t}, \tag{10.46}$$

where $D_i$ are undetermined coefficients. The boundary condition $[u(r = a) = 0]$ requires that

$$J_0(\lambda_i a) = 0, \tag{10.47}$$

which means that parameter $\lambda_i$ are the roots of the Bessel function $J_0(\lambda_i a) = 0$. The initial condition gives

$$\psi(r) = \sum_{i=1}^{\infty} D_i J_0(\lambda_i r). \tag{10.48}$$

Using the basic properties of the Bessel functions

$$\int_0^a r J_0(\lambda_i r) J_0(\lambda_j r) dr = 0, \qquad (\lambda_i \neq \lambda_j), \tag{10.49}$$

and

$$\int_0^a r[J_0(\lambda r)]^2 dr = \frac{a^2}{2} J_1^2(a\lambda), \tag{10.50}$$

$$\int_0^a r J_0(\lambda_i r) dr = \frac{a}{\lambda_i} J_1(a\lambda_i), \tag{10.51}$$

the general solution can be written as

$$u(r,t) = \frac{2}{a^2} \sum_{i=1}^{\infty} \frac{J_0(r\lambda_i)}{J_1^2(a\lambda_i)} e^{-\lambda_i^2 \kappa t} \int_0^a r\psi(r) J_0(r\lambda_i) dr. \tag{10.52}$$

The solution procedure shows that it requires a combination of separation of variables, Bessel functions, and power series.

---

□ ***Example 10.2:*** *If we now want to solve the same diffusion equation in the cylindrical coordinates with slight different boundary conditions:*

$$u(r=a,t)=u_0=const., \qquad u(r,0)=\phi(r),$$

*we have to make a transformation* $u=w+u_0$. *Both* $u$ *and* $w$ *satisfy the same diffusion equation, but now the boundary conditions for* $w$ *become*

$$w(r=a,t)=0, \qquad w(r,0)=\phi(r)-u_0=\psi(r),$$

*which is the problem we have just solved. By substituting* $u=w+u_0$ *and* $\psi(r)=\phi(r)-u_0$ *and using the properties of Bessel functions, we finally obtain*

$$u=u_0[1-\frac{2}{a}\sum_{i=1}^{\infty}\frac{J_0(r\lambda_i)}{\lambda_i J_1(a\lambda_i)}e^{-\lambda_i^2\kappa t}]+\sum_{i=1}^{\infty}\frac{2J_0(r\lambda_i)}{a^2J_1^2(a\lambda_i)}e^{-\lambda_i^2\kappa t}I_{\lambda_i},$$

*where* $I_{\lambda_i}=\int_0^a r\phi(r)J_0(r\lambda_i)dr$. *For a very special case when* $\phi(r)=0$, *we have*

$$u=u_0[1-\sum_{i=1}^{\infty}\frac{2J_0(r\lambda_i)}{a\lambda_i J_1(a\lambda_i)}e^{-\lambda_i^2\kappa t}].$$

□

---

There are other important methods for solving partial differential equations. These include series methods, asymptotic methods, approximate methods, perturbation methods and naturally the numerical methods.

# Chapter 11

# Integral Equations

The calculus of variations is important in many optimization problems and computational sciences, especially the formulation of the finite element methods. On the other hand, integral equations are a different type of equation and they frequently occur in applied mathematics and natural sciences. In this chapter, we will briefly touch these topics.

## 11.1 Calculus of Variations

The main aim of the calculus of variations is to find a function that makes the integral stationary, making the value of the integral a local maximum or minimum. For example, in mechanics we may want to find the shape $y(x)$ of a rope or chain when suspended under its own weight from two fixed points. In this case, the calculus of variations provides a method for finding the function $y(x)$ so that the curve $y(x)$ minimizes the gravitational potential energy of the hanging rope system.

### 11.1.1 Curvature

Before we proceed to the calculus of variations, let us first discuss an important concept, namely the curvature of a curve. In general, a curve $y(x)$ can be described in a parametric form in terms of a vector $\mathbf{r}(s)$ with a parameter $s$ which is the arc length along the curve measured from a fixed point. The curvature $\kappa$ of a curve is defined as the rate at which the unit tangent $\mathbf{t}$ changes with respect to $s$. The change of arc length is

$$\frac{ds}{dx} = \sqrt{1 + (\frac{dy}{dx})^2} = \sqrt{1 + y'^2}. \tag{11.1}$$

We have the curvature

$$\frac{d\mathbf{t}}{ds} = \kappa\, \mathbf{n} = \frac{1}{\rho}\mathbf{n}, \tag{11.2}$$

where $\rho$ is the radius of the curvature, and $\mathbf{n}$ is the principal normal. As the direction of the tangent is defined by the angle $\theta$ made with the $x$-axis by $\mathbf{t}$, we have $\tan\theta = y'$. Hence, the curvature becomes

$$\kappa = \frac{d\theta}{ds} = \frac{d\theta}{dx}\frac{dx}{ds}. \tag{11.3}$$

From $\theta = \tan^{-1} y'(x)$, we have

$$\frac{d\theta}{dx} = [\tan^{-1}(y')]' = \frac{y''}{(1 + y'^2)}. \tag{11.4}$$

Using the expression for $ds/dx$, the curvature can be written in terms of $y(x)$, and we get

$$\kappa = |\frac{d^2\mathbf{r}}{ds^2}| = \frac{y''}{[1 + (y')^2]^{3/2}}. \tag{11.5}$$

### 11.1.2 Euler-Lagrange Equation

Since the calculus of variations is always related to some minimization or maximization, we can in general assume that the integrand $\psi$ of the integral is a function of the shape or curve $y(x)$ (shown in Figure 11.1), its derivative $y'(x)$ and the spatial coordinate $x$ (or time $t$, depending on the context). For the integral

$$I = \int_a^b \psi(x, y, y')dx, \tag{11.6}$$

where $a$ and $b$ are fixed, the aim is to find the solution of the curve $y(x)$ such that it makes the value of $I$ stationary. In this sense, $I[y(x)]$ is a function of the function $y(x)$, and thus it is referred to as the functional.

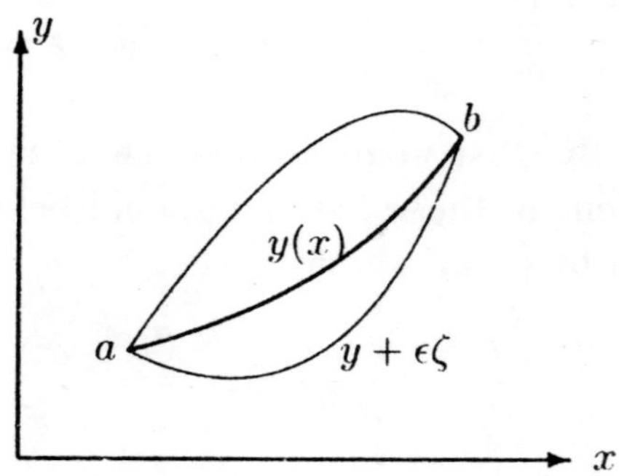


Figure 11.1: Variations in the path $y(x)$.

Here, stationary means that the small change of the first order in $y(x)$ will only lead to the second-order changes in values of $I[y(x)]$, and subsequently, the change $\delta I$ of $I$ should be virtually zero due to the small variation in the function $y(x)$. Translating this into the mathematical language, we suppose that $y(x)$ has a small change of magnitude of $\epsilon$ so that

$$y(x) \to y(x) + \epsilon\zeta(x), \tag{11.7}$$

where $\zeta(x)$ is an arbitrary function. The requirement of $I$ to be stationary means that

$$\delta I = 0, \tag{11.8}$$

or more accurately,

$$\frac{dI}{d\epsilon}\Big|_{\epsilon=0} = 0, \qquad \text{for all} \quad \zeta(x). \tag{11.9}$$

Thus $I$ becomes

$$I(y,\epsilon)=\int_a^b \psi(x,y+\epsilon\zeta,y'+\epsilon\zeta')dx$$

$$=\int_a^b \psi(x,y,y')dx+\int_a^b [\epsilon(\zeta\frac{\partial\psi}{\partial y}+\zeta'\frac{\partial\psi}{\partial y'}]dx+O(\epsilon^2). \quad (11.10)$$

The first derivative of $I$ should be zero, and we have

$$\frac{\delta I}{\delta\epsilon}=\int_a^b [\frac{\partial\psi}{\partial y}\zeta+\frac{\partial\psi}{\partial y'}\zeta']dx=0, \quad (11.11)$$

which is exactly what we mean that the change $\delta I$ (or the first order variation) in the value of $I$ should be zero. Integrating this equation by parts, we have

$$\int_a^b [\frac{\partial\psi}{\partial y}-\frac{d}{dx}\frac{\partial\psi}{\partial y'}]\zeta dx=-[\zeta\frac{\partial\psi}{\partial y'}]\Big|_a^b. \quad (11.12)$$

If we require that $y(a)$ and $y(b)$ are known at the fixed points $x=a$ and $x=b$, then these requirements naturally lead to $\zeta(a)=\zeta(b)=0$. This means that the above right hand side of the equation is zero. That is,

$$\Big[\zeta\frac{\partial\psi}{\partial y'}\Big]_a^b=0, \quad (11.13)$$

which gives

$$\int_a^b \Big[\frac{\partial\psi}{\partial y}-\frac{d}{dx}\frac{\partial\psi}{\partial y'}\Big]\zeta dx=0. \quad (11.14)$$

As this equation holds for all $\zeta(x)$, the integrand must be zero. Therefore, we have the well-known Euler-Lagrange equation

$$\frac{\partial\psi}{\partial y}=\frac{d}{dx}(\frac{\partial\psi}{\partial y'}). \quad (11.15)$$

It is worth pointing out that this equation is very special in the sense that $\psi$ is known and the unknown is $y(x)$. It has many applications in mathematics, natural sciences and engineering.

The simplest and classical example is to find the shortest path on a plane joining two points, say, (0, 0) and (1, 1). We know that the total length along a curve $y(x)$ is

$$L = \int_0^1 \sqrt{1 + y'^2}dx. \tag{11.16}$$

Since $\psi = \sqrt{1 + y'^2}$ does not contain $y$, thus $\frac{\partial \psi}{\partial y} = 0$. From the Euler-Lagrange equation, we have

$$\frac{d}{dx}(\frac{\partial \psi}{\partial y'}) = 0, \tag{11.17}$$

its integration is

$$\frac{\partial \psi}{\partial y'} = \frac{y'}{\sqrt{1 + y'^2}} = A. \tag{11.18}$$

Rearranging it as

$$y'^2 = \frac{A^2}{1 - A^2}, \qquad \text{or} \qquad y' = \frac{A}{\sqrt{1 - A^2}}, \tag{11.19}$$

and integrating again, we have

$$y = kx + c, \qquad k = \frac{A}{\sqrt{1 - A^2}}. \tag{11.20}$$

This is a straight line. That is exactly what we expect from the plane geometry.

---

□ ***Example 11.1:*** *The Euler-Lagrange equation is very general and includes many physical laws if the appropriate form of $\psi$ is used. For a point mass $m$ following under the Earth's gravity $g$, the action (see below) is defined as*

$$\psi = \frac{1}{2}mv^2 - mgy = \frac{1}{2}m(\dot{y})^2 - mgy,$$

*where $y(t)$ is the path, and now $x$ is replaced by $t$. $v = \dot{y}$ is the velocity. The Euler-Lagrange equation becomes*

$$\frac{\partial \psi}{\partial y} = \frac{d}{dt}(\frac{\partial \psi}{\partial v}).$$

*or*

$$-mg = \frac{d}{dt}(mv),$$

*which is essentially the Newton's second law $F = ma$ because the right hand side is the rate of change of the momentum $mv$, and the left hand side is the force.*

□

Well, you may say, this is trivial and there is nothing new about it. This example is indeed too simple. Let us now study a more complicated case so as to demonstrate the wide applications of the Euler-Lagrange equation. In mechanics, there is a Hamilton's principle which states that the configuration of a mechanical system is such that the action integral $I$ of the Lagrangian $\mathcal{L} = T - V$ is stationary with respect to the variations in the path. That is to say that the configuration can be uniquely defined by its coordinates $q_i$ and time $t$, when moving from one configuration at time $t_0$ to another time $t = t*$

$$I = \int_0^{t^*} \mathcal{L}(t, q_i, \dot{q}_i)dt, \qquad i = 1, 2, ..., N, \tag{11.21}$$

where $T$ is the total kinetic energy (usually, a function of $\dot{q}_i$), and $V$ is the potential energy (usually, a function of $q$). Here $\dot{q}_i$ means

$$\dot{q}_i = \frac{\partial q_i}{\partial t}. \tag{11.22}$$

In analytical mechanics and engineering, the Lagrangian $\mathcal{L}$ (=Kinetic energy - Potential energy) is often called the action, thus this principle is also called the principle of least action. The physical configuration or the path of movement follows such a path that makes the action integral stationary.

In the special case, $x \to t$, the Euler-Lagrange equation becomes

$$\frac{\partial \mathcal{L}}{\partial q_i} = \frac{d}{dt}(\frac{\partial \mathcal{L}}{\partial \dot{q}_i}), \tag{11.23}$$

which is the well-known Lagrange's equation. This seems too abstract. Now let us look at a classic example.

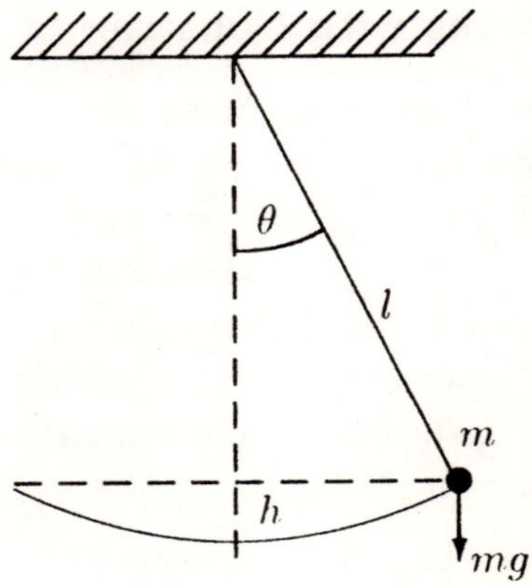


Figure 11.2: A simple pendulum.

---

□ ***Example 11.2:*** For a simple pendulum shown in Figure 11.2, we now try to derive its equation of oscillations. We know the kinetic energy $T$ and the potential energy $V$ are

$$T = \frac{1}{2} m l^2 (\frac{d\theta}{dt})^2 = \frac{1}{2} m l^2 \dot{\theta}^2, \qquad V = mgh = mgl(1 - \cos\theta).$$

Using $\mathcal{L} = T - V$, $q = \theta$ and $\dot{q} = \dot{\theta}$, we have

$$\frac{\partial \mathcal{L}}{\partial \theta} - \frac{d}{dt}(\frac{\partial \mathcal{L}}{\partial \dot{\theta}}) = 0,$$

which becomes

$$-mgl \sin\theta - \frac{d}{dt}(ml^2 \dot{\theta}) = 0.$$

Therefore, we have the pendulum equation

$$\frac{d^2\theta}{dt^2} + \frac{g}{l} \sin\theta = 0.$$

This is a nonlinear equation. If the angle is very small ($\theta \ll 1$), $\sin\theta \approx \theta$, we then have the standard equation for the linear harmonic motion

$$\frac{d^2\theta}{dt^2} + \frac{g}{l}\theta = 0.$$

□

---

### 11.1.3 Variations with Constraints

Although the stationary requirement in the calculus of variations leads to the minimization of the integral itself, there is no additional constraint. In this sense, the calculus of variation discussed up to now is unconstrained. However, sometimes these variations have certain additional constraints, for example, the sliding of a bead on a hanging string. Now we want to make the integral $I$ stationary under another constraint integral $Q$ that is constant. We have

$$I = \int_a^b \psi(x, y, y')dx, \tag{11.24}$$

subjected to the constraint

$$Q = \int_a^b \phi(x, y, y')dx. \tag{11.25}$$

As for most optimization problems under additional constraints, the method of Lagrange multipliers can transform the constrained problem into an unconstrained one by using a combined functional $J = I + \lambda Q$ or

$$J = \int_a^b [\psi + \lambda\phi]dx, \tag{11.26}$$

where $\lambda$ is the undetermined Lagrange multiplier. Replacing $\psi$ by $[\psi + \lambda\phi]$ in the Euler-Lagrange equation or following the same procedure of the derivations, we have

$$[\frac{\partial\psi}{\partial y} - \frac{d}{dx}(\frac{\partial\psi}{\partial y'})] + \lambda[\frac{\partial\phi}{\partial y} - \frac{d}{dx}(\frac{\partial\phi}{\partial y'})] = 0. \tag{11.27}$$

Now we can come back to our example of the hanging rope problem with two fixed points. The total length of the rope is $L$, and it hangs from two fixed points $(-d, 0)$ and $(d, 0)$. From the geometric consideration, it requires that $2d < L$. In order to find the shape of the hanging rope under gravity, we now define its gravitational potential energy $E_p$ as

$$E_p = \int_{-d}^{d} [\rho g y(x) ds] = \rho g \int_{-d}^{d} y\sqrt{1 + y'^2}dx. \tag{11.28}$$

The additional constraint is that the total length of the rope is a constant ($L$). Thus,

$$Q = \int_{-d}^{d} \sqrt{1 + y'^2} dx = L. \tag{11.29}$$

By using the Lagrange multiplier $\lambda$, we have $J = E_p + \lambda Q$, or

$$J = \int_{-d}^{d} [\rho g y + \lambda] \sqrt{1 + y'^2} dx. \tag{11.30}$$

Since $\Psi = [\rho g y + \lambda]\sqrt{1 + y'^2}$ does not contain $x$ explicitly, or $\frac{\partial \Psi}{\partial x} = 0$, then the Euler-Lagrange equation can be reduced into a simpler form in this special case. Using

$$\frac{d\Psi}{dx} = \frac{\partial \Psi}{\partial x} + \frac{\partial \Psi}{\partial y}\frac{dy}{dx} + \frac{\partial \Psi}{\partial y'}\frac{dy'}{dx}$$

$$= 0 + y'\frac{\partial \Psi}{\partial y} + y''\frac{\partial \Psi}{\partial y'}, \tag{11.31}$$

and the Euler-Lagrange equation $\frac{\partial \Psi}{\partial y} = \frac{d}{dx}(\frac{\partial \Psi}{\partial y'})$, we have

$$\frac{d\Psi}{dx} = y'[\frac{d}{dx}(\frac{\partial \Psi}{\partial y'})] + y''\frac{\partial \Psi}{\partial y'} = \frac{d}{dx}[y'\frac{\partial \Psi}{\partial y'}], \tag{11.32}$$

which can again be written as

$$\frac{d}{dx}[\Psi - y'\frac{\partial \Psi}{\partial y'}] = 0. \tag{11.33}$$

The integration of this equation gives

$$\Psi - y'\frac{\partial \Psi}{\partial y'} = A = const. \tag{11.34}$$

Substituting the expression of $\Psi$ into the above equation, the stationary values of $J$ requires

$$\sqrt{1 + y'^2} - \frac{y'^2}{\sqrt{1 + y'^2}} = \frac{A}{\rho g y + \lambda}. \tag{11.35}$$

Multiplying both sides by $\sqrt{1+y'^2}$ and using the substitution $A\cosh\zeta = \rho g y + \lambda$, we have

$$y'^2 = \cosh^2\zeta - 1, \tag{11.36}$$

whose solution is

$$\cosh^{-1}[\frac{\rho g y + \lambda}{A}] = \frac{x\rho g}{A} + K. \tag{11.37}$$

Using the boundary conditions at $x = \pm d$ and the constraint $Q = L$, we have $K = 0$ and implicit equation for $A$

$$\sinh(\frac{\rho g d}{A}) = \frac{\rho g L}{2A}. \tag{11.38}$$

Finally, the curve for the hanging rope becomes the following catenary

$$y(x) = \frac{A}{\rho g}[\cosh(\frac{\rho g x}{A}) - \cosh(\frac{\rho g d}{A})]. \tag{11.39}$$

---

□ ***Example 11.3:*** *For the hanging rope problem, what happens if we only fix one end at $(a, 0)$, while allowing the free end of the hanging rope to slide on a vertical pole? Well, this forms a variation problem with variable end-point(s). We assume that free end is at $(0, y)$ where $y$ acts like a free parameter to be determined. Now the boundary condition at the free end is different. Since the variation of $\delta I = 0$, we have*

$$\delta J = \int_a^b [\frac{\partial \Psi}{\partial y} - \frac{d}{dx}(\frac{\partial \Psi}{\partial y'})]\zeta dx + [\zeta\frac{\partial \Psi}{\partial y'}]_a^b = 0.$$

*As the variation $\zeta$ is now non-zero at the free end point, we then have*

$$\frac{\partial \Psi}{\partial y'} = 0.$$

*From $J = E_p + \lambda Q$, we have $\Psi = (\rho g y + \lambda)\sqrt{1+y'^2}$. Thus, we get*

$$\frac{\partial}{\partial y'}[(\rho g y + \lambda)\sqrt{1+y'^2}] = 0,$$

*or*

$$y'(\rho g y + \lambda)/\sqrt{1 + y'^2} = 0, \qquad \text{or} \qquad y' = 0.$$

*In other words, the slope is zero at the free end.* □

Such a boundary condition of $y' = 0$ has the real physical meaning because any non-zero gradient at the free end would have a non-zero vertical component, thus causing the vertical slip along the rope due to the tension in the rope. The zero-gradient leads to the static equilibrium. Thus, the whole curve of the hanging rope with one free end forms half the catenary.

---

□ ***Example 11.4:*** *Dido's problem concerns the strategy to enclose a maximum area with a fixed length circumference. Legend says that Dido was promised a piece of land on the condition that it was enclosed by an oxhide. She had to cover as much as land as possible using the given oxhide. She cut the oxhide into narrow strips with ends joined, and a whole region of a hill was enclosed.*

*Suppose the total length of the oxhide strip is $L$. The enclosed area $A$ to be maximized is*

$$A = \int_{x_a}^{x_b} y(x)dx,$$

*where $x_a$ and $x_b$ are two end points (of course they can be the same points). We also have the additional constraint*

$$\int_{x_a}^{x_b} \sqrt{1 + y'^2}dx = L = const.$$

*This forms an isoperimetric variation problem. As $L$ is fixed, thus the maximization of $A$ is equivalent to make $I = A + \lambda L$ stationary. That is*

$$I = A + \lambda L = \int_{x_a}^{x_b} [y + \lambda\sqrt{1 + y'^2}]dx.$$

*Using the Euler-Lagrange equation, we have*

$$\frac{\partial I}{\partial y} - \frac{d}{dx}\frac{\partial I}{\partial y'} = 0,$$

*or*

$$\frac{\partial}{\partial y}[y + \lambda\sqrt{1 + y'^2}] + \frac{d}{dx}\frac{\partial}{\partial y'}[y + \lambda\sqrt{1 + y'^2}] = 0,$$

*which becomes*

$$1 - \lambda \frac{d}{dx}\Big(\frac{y'}{\sqrt{1+y'^2}}\Big) = 0.$$

*Integrating it once, we get*

$$\frac{\lambda y'}{\sqrt{1+y'^2}} = x + K,$$

*where $K$ is the integration constant. By rearranging, we have*

$$y' = \pm \frac{x+K}{\sqrt{\lambda^2 - (x+K)^2}}.$$

*Integrating this equation again, we get*

$$y(x) = \mp\sqrt{\lambda^2 - (x+K)^2} + B,$$

*where $B$ is another integration constant. This is equivalent to*

$$(x+K)^2 + (y-B)^2 = \lambda^2,$$

*which is essentially the standard equation for a circle with the centre at $(-K, B)$ and a radius $\lambda$. Therefore, the most area that can be enclosed by a fixed length is a circle.* □

An interesting application is the design of the slides in playgrounds. Suppose we want to design a smooth (frictionless) slide, what is the best curve/shape the slide should take so that a child can slide down in a quickest way? This problem is related to the brachistochrone problem, also called the shortest time problem or steepest descent problem, which initiated the development of the calculus of variations. In 1696, Johann Bernoulli posed a problem to find the curve that minimizes the time for a bead attached to a wire to slide from a point $(0, h)$ to a lower point $(a, 0)$. It was believed that Newton solved it within a few hours after receiving it. From the conservation of energy, we can determine the speed of the bead from the equation $\frac{1}{2}mv^2 + mgy = mgh$, and we have

$$v = \sqrt{2g(h-y)}. \tag{11.40}$$

So the total time taken to travel from $(0, h)$ to $(a, 0)$ is

$$t = \int_0^a \frac{1}{v} ds = \int_0^a \frac{\sqrt{1 + y'^2}}{\sqrt{2g(h - y)}} dx. \tag{11.41}$$

Using the simplified Euler-Lagrange equation (11.34) because the integrand $\Psi = \sqrt{1 + y'^2}/\sqrt{2g(h - y)}$ does not contain $x$ explicitly, we have

$$\sqrt{\frac{(1 + y'^2)}{2g(h - y)}} - y' \frac{\partial}{\partial y'} [\sqrt{\frac{(1 + y'^2)}{2g(h - y)}}] = A. \tag{11.42}$$

By differentiation and some rearrangements, we have

$$y'^2 = \frac{B - h + y}{h - y}, \qquad B = \frac{1}{2gA^2}. \tag{11.43}$$

By changing of variables $\eta = h - y = \frac{B}{2}(1 - \cos\theta)$ and integrating, we have

$$x = \frac{B}{2}[\theta - \sin\theta] + k, \tag{11.44}$$

where $\theta < \pi$ and $k$ is an integration constant. As the curve must pass the point $(0, h)$, we get $k = 0$. So the parametric equations for the curve become

$$x = \frac{B}{2}(\theta - \sin\theta), \qquad y = h - \frac{B}{2}(1 - \cos\theta). \tag{11.45}$$

This is a cycloid, not a straight line, which seems a bit surprising, or at least it is rather counter-intuitive. The bead travels a longer distance, thus has a higher average velocity and subsequently falls quicker than traveling in a straight line.

### 11.1.4 Variations for Multiple Variables

What we have discussed so far mainly concerns the variations in 2-D, and subsequently the variations are in terms $y(x)$ or curves only. What happens if we want to study a surface in the full 3-D configuration? The principle in the previous sections can be extended to any dimensions with multiple variables, however, we will focus on the minimization of a surface here. Suppose we want to study the shape of a soap bubble, the principle of least action leads to the minimal surface problem. The surface integral of a soap bubble should be stationary. Now we assume that the shape of the bubble is $u(x, y)$, then the total surface area is

$$A(u) = \iint_\Omega \Psi dx dy = \iint_\Omega \sqrt{1 + (\frac{\partial u}{\partial x})^2 + (\frac{\partial u}{\partial y})^2} dx dy, \tag{11.46}$$

where

$$\Psi = \sqrt{1 + (\frac{\partial u}{\partial x})^2 + (\frac{\partial u}{\partial y})^2} = \sqrt{1 + u_x^2 + u_y^2}. \tag{11.47}$$

In this case, the extended Euler-Lagrangian equation for two variables $x$ and $y$ becomes

$$\frac{\partial \Psi}{\partial u} - \frac{\partial}{\partial x}(\frac{\partial \Psi}{\partial u_x}) - \frac{\partial}{\partial y}(\frac{\partial \Psi}{\partial u_y}) = 0. \tag{11.48}$$

Substituting $\Psi$ into the above equation and using $\frac{\partial \Psi}{\partial u} = \Psi_u = 0$ since $\Psi$ does not contain $u$ explicitly, we get

$$-\frac{\partial}{\partial x}[\frac{1}{\Psi}\frac{\partial u}{\partial x}] - \frac{\partial}{\partial y}[\frac{1}{\Psi}\frac{\partial u}{\partial y}] = 0, \tag{11.49}$$

or

$$(1 + u_y^2)u_{xx} - 2u_x u_y + (1 + u_x^2)u_{yy} = 0. \tag{11.50}$$

This is a nonlinear equation and its solution is out of the scope of this book. This nonlinear equation has been one of the active research topics for more than a century. It has been proved that the fundamental solution to this equation is a sphere, and in fact we know that all bubbles are spherical. For some problems, we can approximately assume that $u_x$ and $u_y$ are small, thus the above equation becomes Laplace's equation

$$u_{xx} + u_{yy} = 0. \tag{11.51}$$

The calculus of variations has many applications. The other classical examples include Fermat's principle in optics, Sturm-Liouville problem, surface shape minimization, the action principle. and of course the finite element analysis.

## 11.2 Integral Equations

From the calculus of variations, we know that the unknown $y(x)$ to be optimized is inside the integrand of $I$. In certain sense, this is an integral equation. In fact, many physical processes and laws of conservation are expressed in terms of integral forms rather than their differentiation counterparts. Naturally, one of the ways of constructing an integral equation is to integrate from a differential equation. Integral equations are much more complicated compared with the differential equations. There is no universal solution technique for nonlinear equations, even the numerical simulations are usually not straightforward. Thus, we will mainly focus on the simplest types of integral equations.

### 11.2.1 Linear Integral Equations

#### Fredholm Integral Equations

A linear integral equation for $y(x)$ can be written in the following generic form

$$u(x) + \lambda \int_a^b K(x, \eta) y(\eta) d\eta = v(x) y(x), \tag{11.52}$$

where $K(x, \eta)$ is referred to as the kernel of the integral equation. The parameter $\lambda$ is a known constant. If the function $u(x) = 0$, the equation is then called homogeneous. If $u(x) \neq 0$, the equation is inhomogeneous.

If the function $v(x) = 0$, then the unknown $y(x)$ appears only once in the integral equation, and it is under the integral sign only. This is called the linear integral equation of the first kind

$$u(x) + \lambda \int_a^b K(x, \eta) y(\eta) d\eta = 0. \qquad (11.53)$$

On the other hand, if $v(x) = 1$, equation (11.52) becomes the integral equation of the second kind

$$u(x) + \lambda \int_a^b K(x, \eta) y(\eta) d\eta = y(x). \qquad (11.54)$$

An integral equation with the fixed integration limits $a$ and $b$, is called a Fredholm equation. If the upper integration limit $b$ is not fixed, then the equation becomes a Volterra equation. The integral equation becomes singular and at least one of its integration limits approaches infinite.

### Volterra Integral Equation

In general, the Volterra integral equation can be written as

$$u(x) + \lambda \int_a^x K(x, \eta) y(\eta) d\eta = v(x) y(x). \qquad (11.55)$$

The first kind [or $v(x) = 0$] and second kind [or $v(x) = 1$] are defined in the similar manner.

The kernel is said to be separable or degenerate if it can be written in the finite sum form

$$K(x, \eta) = \sum_{i=1}^{N} f_i(x) g_i(\eta), \qquad (11.56)$$

where $f_i(x)$ and $g_i(\eta)$ are functions of $x$ and $\eta$, respectively. A kernel is called a displacement kernel if it can be written as a function of the difference $(x - \eta)$ of its two arguments

$$K(x, \eta) = K(x - \eta). \qquad (11.57)$$

## 11.3 Solution of Integral Equations

Most integral equations do not have closed-form solutions. For linear integral equations, the closed-form solutions are only possible for the special cases of separable and displacement kernels.

### 11.3.1 Separable Kernels

For a Fredholm integral equation of the second kind with separable kernels, we can substitute the kernel (11.56) into the equation and we have

$$u(x) + \lambda \int_a^b \sum_{i=1}^{N} f_i(x) g_i(\eta) d\eta = y(x), \tag{11.58}$$

which becomes

$$u(x) + \lambda \sum_{i=1}^{N} f_i(x) \int_a^b g_i(\eta) d\eta = y(x). \tag{11.59}$$

Because the integration limits are fixed, the integrals over $\eta$ should be constants that are to be determined. By defining

$$\alpha_i = \int_a^b g_i(\eta) y(\eta) d\eta, \tag{11.60}$$

we now have the solution in the form

$$y(x) = u(x) + \lambda \sum_{i=1}^{N} \alpha_i f_i(x), \tag{11.61}$$

where the $N$ coefficients $\alpha_i$ are determined by

$$\alpha_i = \int_a^b g_i(\eta) u(\eta) d\eta + \lambda \sum_{i=1}^{N} \int_a^b [\alpha_i f_i(\eta) g_i(\eta)] d\eta. \tag{11.62}$$

for $i = 1, 2, ..., N$. Only for a few special cases, these coefficients can be written as simple explicit expressions.

### 11.3.2 Di: placement Kernels

For a singular integral equation with a displacement kernel, the equation can be solved by Fourier transforms if both integration limits of the integral are infinite. In this case, we have

$$u(x) + \lambda \int_{-\infty}^{\infty} K(x-\eta) y(\eta) d\eta = y(\eta). \tag{11.63}$$

Using the Fourier transforms and the convolution theorem, we have

$$\bar{U}(\omega) + \lambda\sqrt{2\pi}\bar{K}(\omega)\bar{Y}(\omega) = \bar{Y}(\omega), \tag{11.64}$$

which is an algebraic equation for $\bar{Y}(\omega)$. Its solution is simply

$$\bar{Y}(\omega) = \frac{\bar{U}(\omega)}{1 - \lambda\sqrt{2\pi}\bar{K}(\omega)}. \tag{11.65}$$

The solution $y(x)$ can be obtained using the inverse Fourier transform

$$y(x) = \frac{1}{\sqrt{2\pi}} \int_{-\infty}^{\infty} \frac{\bar{U}(\omega)}{[1 - \lambda\sqrt{2\pi}\bar{K}(\omega)]} e^{i\omega x} d\omega. \tag{11.66}$$

### 11.3.3 Volterra Equation

A Volterra equation with separable kernels may be solved by transforming into a differential equation via direct differentiation. In the case of a simple degenerate kernel

$$K(x, \eta) = f(x) g(\eta), \tag{11.67}$$

we have

$$y(x) = u(x) + \lambda \int_0^x f(x) g(\eta) y(\eta) d\eta, \tag{11.68}$$

which becomes

$$y(x) = u(x) + \lambda f(x) \int_0^x g(\eta) y(\eta) d\eta. \tag{11.69}$$

If $f(x) \neq 0$, it can be written as

$$\frac{y(x)}{f(x)} = \frac{u(x)}{f(x)} + \lambda \int_0^x g(\eta)y(\eta)d\eta. \qquad (11.70)$$

Putting $\phi(x) = u(x)/f(x)$ and differentiating it, we have

$$[\frac{y(x)}{f(x)}]' = \phi'(x) + \lambda g(x)y(x). \qquad (11.71)$$

By letting $\Psi = y(x)/f(x)$. we have

$$\Psi'(x) - \lambda f(x)g(x)\Psi(x) = \phi'(x), \qquad (11.72)$$

which is a first-order ordinary differential equation for $\Psi(x)$. This is equivalent to the standard form

$$\Psi' + P(x)\Psi = Q(x), \qquad (11.73)$$

and

$$P(x) = -\lambda f(x)g(x), \qquad Q(x) = [\frac{u(x)}{f(x)}]'. \qquad (11.74)$$

We can use the standard technique by multiplying the integrating factor $\exp[\int P(x)dx]$ to obtain the solution. We get

$$y(x) = f(x)[e^{-\int P(x)dx}]\{\int [Q(x)e^{\int P(x)dx}]dx\}. \qquad (11.75)$$

With appropriate boundary conditions, the exact form of the solution can be obtained.

---

□ ***Example 11.5:*** *Let us try to solve the integral equation of Volterra type*

$$y(x) = e^x + \int_0^x e^x \sin(\zeta)y(\zeta)d\zeta.$$

*First, we divide both sides by* $e^x$, *we get*

$$\frac{y(x)}{e^x} = 1 + \int_0^x \sin(\zeta)y(\zeta)d\zeta,$$

*whose differentiation with respect to $x$ leads to*

$$[\frac{y(x)}{e^x}]' = y(x)\sin(x),$$

*or*

$$\frac{1}{e^x}y'(x) - y(x)e^{-x} = y(x)\sin(x).$$

*Divide both sides by $y(x)$ and using $[\ln y(x)]' = y'(x)/y(x)$, we have*

$$[\ln y(x)]' = e^x \sin x + 1.$$

*By direct integration, we have*

$$\ln y(x) = x - \frac{1}{2}e^x \cos x + \frac{1}{2}e^x \sin x.$$

*Thus, we finally obtain*

$$y(x) = \exp[x - \frac{e^x}{2}(\cos x - \sin x)].$$

□

There are other methods and techniques of solving integral equations such as the operator method, series method and the Fredholm theory. However, most integral equations do not have closed-form solutions. In this case, numerical methods are the best alternative.

# Chapter 12

# Tensor Analysis

Many physical quantities such as stresses and strains are tensors. Vectors are essentially first-order tensors. Tensors are the extension of vectors, and they can have any number of dimensions and any orders, though most commonly used tensors are second order tensors.

## 12.1 Notations

In tensor analysis, the Einstein summation convention or Einstein notations [1] and notations for subscripts are widely used. Any lowercase subscript that appears exactly twice in any term of an expression means that sum is over its all possible values of the subscript. For example, in the three-dimensional case, we have

$$a_i x_i \equiv \sum_{i=1}^{3} a_i x_i = a_1 x_1 + a_2 x_2 + a_3 x_3. \tag{12.1}$$

$$A_{ij} B_{jk} \equiv \sum_{j=1}^{3} A_{ij} B_{jk} = A_{i1} B_{1k} + A_{i2} B_{2k} + A_{i3} B_{3k}. \tag{12.2}$$

[1] This notation convention was introduced by Albert Einstein in 1916 when formulating the theory of General Relativity.

$$\frac{\partial u_i}{\partial x_i} \equiv \nabla \cdot \mathbf{u} = \frac{\partial u_1}{\partial x_1} + \frac{\partial u_2}{\partial x_2} + \frac{\partial u_3}{\partial x_3}. \tag{12.3}$$

The Kronecker delta $\delta_{ij}$ which is a unity tensor (like the unity matrix $\mathbf{I}$ in matrix analysis), is defined as

$$\delta_{ij} = \begin{cases} 1 \ (\text{if} \ \ i = j), \\ 0 \ (\text{if} \ \ i \neq j). \end{cases} \tag{12.4}$$

For a tensor with three subscripts similar to $\delta_{ij}$, the Levi-Civita symbol or tensor is defined as

$$\epsilon_{ijk} = \begin{cases} +1 \ \text{if (i, j, k) is an even permutation of } (1,2,3), \\ -1 \ \text{if (i, j, k) is an odd permutation of } (1,2,3), \\ 0 \qquad \text{(otherwise)}. \end{cases} \tag{12.5}$$

The tensors $\delta_{ij}$ and $\epsilon_{ijk}$ are related by

$$\epsilon_{ijk}\epsilon_{kpq} = \delta_{ip}\delta_{jq} - \delta_{iq}\delta_{jp}. \tag{12.6}$$

Using the summation conventions, the matrix equation

$$\mathbf{Ax} = \mathbf{b}, \tag{12.7}$$

can alternatively be written as

$$A_{ij}x_j = b_i, \qquad (i = 1, 2, ..., n). \tag{12.8}$$

## 12.2 Tensors

When changing the bases from the standard Cartesian $\mathbf{e_1} = \mathbf{i}$, $\mathbf{e_2} = \mathbf{j}$, $\mathbf{e_3} = \mathbf{k}$ to a new set of bases $\mathbf{e}'_1, \mathbf{e}'_2, \mathbf{e}'_3$, a position vector $\mathbf{x} = (x_1, x_2, x_3)$ in the old bases is related to the new vector $\mathbf{x}' = (x'_1, x'_2, x'_3)$ in the new bases by a coefficient matrix $S_{ij}$. $S_{ij}$ can be the rotation, translation, enlargement or any of their combinations. For example, the matrix for a simple rotation, with an angle of $\theta$ around a fixed axis, $S_{ij}$ becomes

$$S_{ij} = \begin{pmatrix} \cos\theta & \sin\theta & 0 \\ -\sin\theta & \cos\theta & 0 \\ 0 & 0 & 1 \end{pmatrix} \tag{12.9}$$

The orthogonality of $S_{ij}$ requires that $\mathbf{SS}^\mathbf{T} = \mathbf{S}^\mathbf{T}\mathbf{S} = \mathbf{I}$ or

$$S_{ij}S_{jk} = \delta_{ik}, \qquad S_{ki}S_{kj} = \delta_{ij}. \tag{12.10}$$

If the components $u_i$ of any variable $\mathbf{u}$ are transformed to the components $u'_i$ in the new bases in the same manner as

$$u'_i = S_{ij}u_j, \qquad u_i = S_{ji}u'_j, \tag{12.11}$$

then $u_i(i = 1, 2, 3)$ are said to form a first-order Cartesian tensor (or vector in this case). If components of a variable such as strains $\sigma_{ij}$ are transformed as

$$\sigma'_{ij} = S_{ip}S_{jq}\sigma_{pq}, \qquad \sigma_{ij} = S_{pi}S_{qj}\sigma'_{pq}. \tag{12.12}$$

we say these components form a second-order tensor.

The order of a tensor is also called its rank. Scalars have rank zero, vectors have rank 1, and second-order tensors have rank 2. In engineering and computing, the rank is associated with the number of indices to describe a tensor in terms of a multidimensional array. In this sense, a second-order tensor is equivalent to a two-dimensional array or a matrix.

In a similar fashion, higher-order tensors can be defined, and for each order increase, then there is one $S_{ij}$ extra in the product for transforming, but no subscripts are allowed to appear more than twice

$$\tau'_{ij...k} = S_{ip}S_{jq}...S_{kr}\tau_{pq...r}. \tag{12.13}$$

and

$$\tau_{ij...k} = S_{pi}S_{qj}....S_{rk}\tau'_{pq...r}. \tag{12.14}$$

## 12.3 Tensor Analysis

### Tensors in Cartesian Coordinates

One of the main advantages of tensors is that a tensor is independent of any chosen frame of reference. Therefore, any physical laws or equations that are formulated in terms of tensors should be independent of frame of reference. For example, the stress -strain relation in linear elasticity is independent of frame of reference (for details see next chapter)

$$\sigma_{ij} = 2\mu\varepsilon_{ij} + \lambda\varepsilon_{kk}\delta_{ij}, \tag{12.15}$$

where $\sigma_{ij}$ and $\varepsilon_{ij}$ are the stress tensor and strain tensor, respectively. $\mu$ and $\lambda$ are Lamé constants.

In the similar way as multi-dimensional arrays or matrices, two tensors can be added or subtracted component-by-component if and only if they are the tensors of the same order. For second-order tensors, a tensor $\tau_{ij}$ is said to be symmetric if $\tau_{ij} = \tau_{ji}$, and antisymmetric if $\tau_{ij} = -\tau_{ji}$. An interesting property of a tensor $\tau_{ij}$ is that it can always be written as a sum of a symmetric tensor and an antisymmetric tensor

$$\tau_{ij} = \frac{1}{2}(\tau_{ij} + \tau_{ji})[\text{sym.}] + \frac{1}{2}(\tau_{ij} - \tau_{ji})[\text{antisym.}]. \tag{12.16}$$

All the formulas in vector analysis can be rewritten in the tensor forms using the summation convention and notations

$$\mathbf{u} \cdot \mathbf{v} = u_i v_i = \delta_{ij} u_i v_j, \tag{12.17}$$

$$\nabla^2 \psi = \frac{\partial^2 \psi}{\partial x_i \partial x_i} = \delta_{ij} \frac{\partial \psi}{\partial x_i \partial x_j}, \tag{12.18}$$

$$\nabla \times (\nabla \times \mathbf{u})_i = \epsilon_{ijk}\epsilon_{kpq} \frac{\partial u_q}{\partial x_j x_p}. \tag{12.19}$$

Similarly, the divergence theorem can be rewritten as the following form

$$\int_V \frac{\partial u_i}{\partial x_i} dV = \oint_S u_i n_i dS. \tag{12.20}$$

The tensor forms are sometimes useful to the proof of the complex relationship among vectors and tensors. They also become handy for the implementation of numerical algorithms.

Using the tensor notations, we have the identity $\delta_{ij}A_j = A_i$. The cross product $\mathbf{A} \times \mathbf{B}$ can be expressed as

$$C_i = \epsilon_{ijk} A_j B_k. \tag{12.21}$$

Another way of denoting the derivatives of tensors $\sigma$ and $\mathbf{v}$ is to use following notations

$$\sigma_{,i} \equiv \frac{\partial \sigma}{\partial x_i}, \qquad v_{i,jk} = \frac{\partial^2 v_i}{\partial x_j \partial x_k}, \tag{12.22}$$

where the index of the spatial component $x_i$ is denoted by a comma to avoid any potential confusion with other indices. With these notations, the important operators involving the $\nabla$-operator can be written as

$$\text{grad}\phi = \phi, i = \nabla\phi. \tag{12.23}$$

$$\nabla^2\phi = \phi_{,ii} = \Delta\phi, \tag{12.24}$$

and

$$\nabla.\mathbf{v} = v_{i,i} = \frac{\partial^2 u_1}{\partial {x_1}^2} + \frac{\partial^2 u_2}{\partial {x_2}^2} + \frac{\partial^2 u_3}{\partial {x_3}^2}. \tag{12.25}$$

A very special case is that

$$x_{i,j} = \delta_{ij}. \tag{12.26}$$

---

□ ***Example 12.1:*** *Let us now use the tensor notations to prove* $\mathbf{a} \times (\mathbf{b} \times \mathbf{c}) = \mathbf{b}(\mathbf{a} \cdot \mathbf{c}) - \mathbf{c}(\mathbf{a} \cdot \mathbf{b})$. *From above expressions for cross products, we know that*

$$\mathbf{a} \times (\mathbf{b} \times \mathbf{c}) = \epsilon_{ijk} a_j (\epsilon_{kpq} b_p c_q) = \epsilon_{ijk}\epsilon_{kpq} a_j b_p c_q$$

$$= (\delta_{ip}\delta_{jq} - \delta_{iq}\delta_{jp}) a_j b_p c_q = \delta_{ip} b_p (\delta_{jq} a_j c_q) - \delta_{iq} c_q (\delta_{jp} a_j b_p).$$

*Using* $\delta_{ip} b_p = \mathbf{b}$ *and* $\delta_{iq} c_q = \mathbf{c}$, *we have*

$$\mathbf{a}(\mathbf{b} \times \mathbf{c}) = \mathbf{b}(\delta_{jq} a_j c_q) - \mathbf{c}(\delta_{jp} a_j b_p).$$

*By renaming the indices* ($j \to i$ *and* $q \to j$) *so that* $\delta_{jq} a_j c_q = \delta_{ij} a_i c_j = \boldsymbol{a} \cdot \mathbf{c}$ *and* $\delta_{jp} a_j b_p = \boldsymbol{a} \cdot \boldsymbol{b}$, *we finally obtain*

$$\mathbf{a} \times (\mathbf{b} \times \mathbf{c}) = \mathbf{b}(\mathbf{a} \cdot \mathbf{c}) - \mathbf{c}(\mathbf{a} \cdot \mathbf{b}).$$

□

---

## Tensors in Non-Cartesian Coordinates

The tensors we have discussed so far are expressed in Cartesian coordinates. In non-Cartesian coordinates, they are more complicated. Tensor analysis is very important in theoretical physics and differential geometry where formal mathematical theory is required. In fact, many books on tensor analysis use a modern approach in terms of tensor duality, covariance and contravariance concepts. In the simplest term, a tensor such as a vector $\mathbf{v}$ can be expressed as the sum of its components multiplying by the basis vectors

$$\mathbf{v} = v^i \boldsymbol{e}_i = v_i \boldsymbol{e}^i, \tag{12.27}$$

where $v_i (i = 1, 2, 3)$ are called the covariant components of $\mathbf{v}$ in the contravariant basis vectors $\boldsymbol{e}^i$, while $v^i (i = 1, 2, 3)$ are called the contravariant components of $\mathbf{v}$ in the covariant basis vectors $\boldsymbol{e}_i$. For the curvilinear coordinates $(q_1, q_2, q_3)$ at any point $P$ on a position vector $\mathbf{r}(q_1, q_2, q_3)$, the basic vectors are given by

$$\boldsymbol{e}_i = \frac{\partial \mathbf{r}}{\partial q_i}, \qquad \boldsymbol{e}^i = \nabla q_i, \qquad (i = 1, 2, 3), \tag{12.28}$$

where $\boldsymbol{e}_i$ and $\boldsymbol{e}^i$ are reciprocal systems of vectors, and $\boldsymbol{e}^i \cdot \boldsymbol{e}_j = \delta^i_j$, where $\delta^i_j$ acts in the similar way as $\delta_{ij}$. In many books on tensor analysis, $\boldsymbol{e}^i$ is also written as $\boldsymbol{\epsilon}_i \equiv \boldsymbol{e}^i$.

Furthermore, the tensor product (also called outer product) of two tensors is rather complicated. For example, the tensor product $\mathbf{u} \otimes \mathbf{v} = \mathbf{u}\mathbf{v}^{\mathbf{T}}$ of two vectors $\mathbf{u}$ and $\mathbf{v}$ is given by

$$\mathbf{u} \otimes \mathbf{v} = \begin{pmatrix} u_1 \\ u_2 \\ u_3 \end{pmatrix} \otimes (v_1 \ v_2 \ v_3) = \begin{pmatrix} u_1 v_1 & u_1 v_2 & u_1 v_3 \\ u_2 v_1 & u_2 v_2 & u_2 v_3 \\ u_3 v_1 & u_3 v_2 & u_3 v_3 \end{pmatrix}. \tag{12.29}$$

Using these basis vectors, we can write a second-order tensor in terms of covariant components $\sigma_{ij}$ and contravariant components $\sigma^{ij}$ as

$$\boldsymbol{\sigma} = \sigma_{ij} \, \boldsymbol{e}^i \otimes \boldsymbol{e}^j = \sigma^{ij} \boldsymbol{e}_i \otimes \boldsymbol{e}_j. \tag{12.30}$$

In a given frame of reference, the fundamental metric tensor is defined by

$$g_{ij} = \boldsymbol{e}_i \cdot \boldsymbol{e}_j. \tag{12.31}$$

This tensor is always symmetric $g_{ij} = g_{ji}$, its determinant $g \equiv |g_{ij}| = \det(g_{ij})$ is related to the Jacobian $J = \sqrt{g}$.

The derivatives in non-Cartesian coordinates are far more complicated, and they usually involve the Christoffel coefficients $\Gamma^k_{ij} = \boldsymbol{e}^k \cdot \frac{\partial \boldsymbol{e}_i}{\partial q_j}$. For example, the divergence of a vector is defined by the covariant differentiation $u^j_{;j}$ or $\nabla \cdot \boldsymbol{u} = u^j_{;j} = \frac{\partial u^j}{\partial q_j} + \Gamma^j_{ij} u^i = u^j_{,j} + \Gamma^j_{ij} u^i$. It is worth pointing out that $\boldsymbol{e}_i$ and $\boldsymbol{e}^i$ become identical and $\Gamma^k_{ij} = 0$ in Cartesian coordinates, and it is therefore not necessary to distinguish the contravariant and covariant vectors and components.

Mathematically speaking, the formal approach is preferred. In engineering mathematics, however, the simple formulation in terms of multidimensional arrays in Cartesian coordinates is a more convenient approach, especially from the computational point of view. That is why we have used an over-simplified approach here. In the next chapter, we will study the theory of linear elasticity as an application of tensor analysis.

# Chapter 13

# Elasticity

## 13.1 Hooke's Law and Elasticity

The basic Hooke's law of elasticity concerns an elastic body such as a spring, and it states that the extension $x$ is proportional to the load $F$, that is

$$F = kx, \tag{13.1}$$

where $k$ the spring constant. However, this equation only works for 1-D deformations. For a bar of uniform cross-section with a length $L$ and a cross section area $A$, it is more convenient to use strain $\varepsilon$ and stress $\sigma$. The stress and strain are defined by

$$\sigma = \frac{F}{A}, \qquad \varepsilon = \frac{\Delta L}{L}, \tag{13.2}$$

where $\Delta L$ is the extension. The unit of stress is $N/m^2$, while the strain is dimensionless, though it is conventionally expressed in m/m or % (percentage) in engineering. For the elastic bar, the stress-strain relationship is

$$\sigma = E\varepsilon, \tag{13.3}$$

where $E$ is the Young's modulus of elasticity. Written in terms $F$ and $x = \Delta L$, we have

$$F = \frac{EA}{L}\Delta L = kx, \qquad k = \frac{EA}{L}. \tag{13.4}$$

where $k$ is the equivalent spring constant for the bar. This equation is still only valid for any unidirectional compression or extension. For the 2-D and 3-D deformation, we need to generalize Hooke's law. For the general stress tensor (also called Cauchy stress tensor)

$$\boldsymbol{\sigma} = \begin{pmatrix} \sigma_{xx} & \sigma_{xy} & \sigma_{xz} \\ \sigma_{yx} & \sigma_{yy} & \sigma_{yz} \\ \sigma_{zx} & \sigma_{zy} & \sigma_{zz} \end{pmatrix} = \begin{pmatrix} \sigma_{11} & \sigma_{12} & \sigma_{13} \\ \sigma_{21} & \sigma_{22} & \sigma_{23} \\ \sigma_{31} & \sigma_{32} & \sigma_{33} \end{pmatrix}, \tag{13.5}$$

and strain tensor

$$\boldsymbol{\varepsilon} = \begin{pmatrix} \varepsilon_{xx} & \varepsilon_{xy} & \varepsilon_{xz} \\ \varepsilon_{yx} & \varepsilon_{yy} & \varepsilon_{yz} \\ \varepsilon_{zx} & \varepsilon_{zy} & \varepsilon_{zz} \end{pmatrix} = \begin{pmatrix} \varepsilon_{11} & \varepsilon_{12} & \varepsilon_{13} \\ \varepsilon_{21} & \varepsilon_{22} & \varepsilon_{23} \\ \varepsilon_{31} & \varepsilon_{32} & \varepsilon_{33} \end{pmatrix}, \tag{13.6}$$

it can be proved later that these tensors are symmetric, that is $\boldsymbol{\sigma} = \boldsymbol{\sigma}^T$ and $\boldsymbol{\varepsilon} = \boldsymbol{\varepsilon}^T$, which leads to

$$\sigma_{xy} = \sigma_{yx}, \quad \sigma_{xz} = \sigma_{zx}, \quad \sigma_{yz} = \sigma_{zy}, \tag{13.7}$$

and

$$\varepsilon_{xy} = \varepsilon_{yx}, \quad \varepsilon_{xz} = \varepsilon_{zx}, \quad \varepsilon_{yz} = \varepsilon_{zy}. \tag{13.8}$$

Therefore, we only have 6 independent components or unknowns for stresses and 6 unknown strain components.

The strain tensor is defined by the displacement $\mathbf{u}^T = (u_1, u_2, u_3)$

$$\varepsilon_{ij} = \frac{1}{2}\left(\frac{\partial u_i}{\partial x_j} + \frac{\partial u_j}{\partial x_i}\right), \tag{13.9}$$

where $x_1 = x$, $x_2 = y$, and $x_3 = z$. Sometimes, it is useful to write

$$\boldsymbol{\varepsilon} = \frac{1}{2}(\nabla \boldsymbol{u} + \nabla \boldsymbol{u}^T). \tag{13.10}$$

The generalized Hooke's law can be written as

$$\varepsilon_{xx} = \frac{1}{E}[\sigma_{xx} - \nu(\sigma_{yy} + \sigma_{zz})], \tag{13.11}$$

$$\varepsilon_{yy} = \frac{1}{E}[\sigma_{yy} - \nu(\sigma_{xx} + \sigma_{zz})], \tag{13.12}$$

$$\varepsilon_{zz} = \frac{1}{E}[\sigma_{zz} - \nu(\sigma_{xx} + \sigma_{yy})], \tag{13.13}$$

$$\varepsilon_{xy} = \frac{1+\nu}{E}\sigma_{xy}, \tag{13.14}$$

$$\varepsilon_{xz} = \frac{1+\nu}{E}\sigma_{xz}, \tag{13.15}$$

$$\varepsilon_{yz} = \frac{1+\nu}{E}\sigma_{yz}, \tag{13.16}$$

where $\nu$ is the Poisson's ratio, and it measures the tendency of extension in transverse directions (say, $x$ and $y$) when the elastic body is stretched in one direction (say, $z$). It can be defined as the ratio of the transverse contract strain (normal to the applied load) to the axial strain in a stretched cylindrical bar in the direction of the applied force. For a perfectly incompressible material, $\nu = 0.5$, and $\nu = 0 \sim 0.5$ for most common materials. For example, steels have $\nu = 0.25 \sim 0.3$. Some auxetic material such as polymer foams or anti-rubbers have a negative Poisson's ratio $\nu < 0$.

This generalized Hooke's law can concisely be written as

$$\varepsilon_{ij} = \frac{1+\nu}{E}\sigma_{ij} - \frac{\nu}{E}\sigma_{kk}\delta_{ij}, \tag{13.17}$$

where we have used the Einstein's summation convention $\sigma_{kk} = \sigma_{xx}+\sigma_{yy}+\sigma_{zz}$. Another related quantity is the pressure, which is defined by

$$p = -\frac{1}{3}\sigma_{kk} = -\frac{\sigma_{xx} + \sigma_{yy} + \sigma_{zz}}{3}. \tag{13.18}$$

The negative sign comes from the conventions that a positive normal stress results in tension, and negative one in compression, while the positive pressure acts in compression. Sometimes, it is more convenient to express the stress tensor in terms of pressure and devitoric stress tensor $s_{ij}$

$$\sigma_{ij} = -p\delta_{ij} + s_{ij}. \tag{13.19}$$

If we want to invert equation (13.17), we have first express $\sigma_{kk}$ in terms of $\varepsilon_{kk}$ so that the right hand side of the new expression does not contain the stress $\sigma_{kk}$. By contraction using $j \to i$, we have

$$\varepsilon_{ii} = \frac{1+\nu}{E}\sigma_{ii} - \frac{\nu}{E}\sigma_{kk}\delta_{ii} = \frac{1-2\nu}{E}\sigma_{ii}, \tag{13.20}$$

where we have used $\delta_{ii} = \delta_{11} + \delta_{22} + \delta_{33} = 1 + 1 + 1 = 3$ and $\sigma_{ii} = \sigma_{kk}$. In engineering, the quantity

$$\varepsilon_{kk} = \varepsilon_{xx} + \varepsilon_{yy} + \varepsilon_{zz} = \frac{\partial^2 u_1}{\partial x^2} + \frac{\partial^2 u_2}{\partial y^2} + \frac{\partial^2 u_3}{\partial z^2} = \nabla \cdot \boldsymbol{u}, \tag{13.21}$$

means the fractional change in volume, known as the dilation. This gives that

$$\sigma_{ii} = \sigma_{kk} = \frac{E}{1-2\nu}\varepsilon_{kk}. \tag{13.22}$$

Substituting it into equation (13.17), we have

$$\varepsilon_{ij} = \frac{1+\nu}{E}\sigma_{ij} - \frac{\nu}{E}(\frac{E}{1-2\nu}\varepsilon_{kk})\delta_{ij}, \tag{13.23}$$

or after some rearrangement

$$\frac{1+\nu}{E}\sigma_{ij} = \varepsilon_{ij} + \frac{\nu}{1-2\nu}\varepsilon_{kk}\delta_{ij}, \tag{13.24}$$

which can be written as

$$\sigma_{ij} = 2G\varepsilon_{ij} + \lambda\varepsilon_{kk}\delta_{ij}, \tag{13.25}$$

where $\mu$ and $\lambda$ are Lamé constants. They are

$$G = \mu = \frac{E}{2(1+\nu)}, \qquad \lambda = \frac{\nu E}{(1+\nu)(1-2\nu)}. \tag{13.26}$$

This stress-strain relationship can also be written as

$$\boldsymbol{\sigma} = 2G\boldsymbol{\varepsilon} + \lambda(\nabla \cdot \boldsymbol{u})\boldsymbol{\delta}. \tag{13.27}$$

In engineering, $G = \mu$ is called the shear modulus, while $K = \frac{E}{3(1-2\nu)}$ is called the bulk modulus which is the ratio of pressure $-p$ to the volume change $\Delta V$.

## 13.2 Maxwell's Reciprocal Theorem

For an elastostatic problem, the balance of force leads to

$$\nabla \cdot \boldsymbol{\sigma} + \mathbf{b} = 0, \tag{13.28}$$

where $\mathbf{b}$ is the body force or force per unit volume. For a small cube volume element, the total body force $df_i$ along the $x_i$-axis is

$$df_i = -db_i = \frac{\partial \sigma_{ij}}{\partial x_j} dV, \tag{13.29}$$

here we have used the index summation conventions. Similarly, the total force along the $j$-axis is

$$df_j = \frac{\partial \sigma_{ji}}{\partial x_i} dV. \tag{13.30}$$

Since there is no relative rotation of the cube element because the cube element must be at rotational equilibrium, thus the result moment must be zero. Taking the moment of the two force components about any point (say, a corner of the cube). This leads to

$$\sigma_{ij} = \sigma_{ji}, \tag{13.31}$$

which means the stress tensor is symmetric. This is in fact the compatibility condition for stresses.

Alternatively, consider a cube element at rotational equilibrium with a volume $dV = \delta x \delta y \delta z$, and the dimensions of the elements are $\delta x$, $\delta y$ and $\delta z$, respectively. We consider the forces along the four faces that are parallel to $z$-axis (shown in Figure 13.1). If we take the moment about a line which is parallel to $z$-axis and goes through one corner point, we only have to consider the two faces that are far from this point because the two faces through this point do not contribute to the moment. The forces on the two faces are $\sigma_{xy} dA_1 dy$ and $\sigma_{yx} dA_2 dx$ where the surface areas are $dA_1 = \delta x \delta z$ and $dA_2 = \delta y \delta z$. The total moment about point $D$ is

$$\sigma_{xy} dV - \sigma_{yx} dV = 0. \tag{13.32}$$

Thus, $\sigma_{xy} = \sigma_{yx}$. Similar arguments for other faces along $x$- and $y$-directions about any other points and we have $\sigma_{ij} = \sigma_{ji}$.

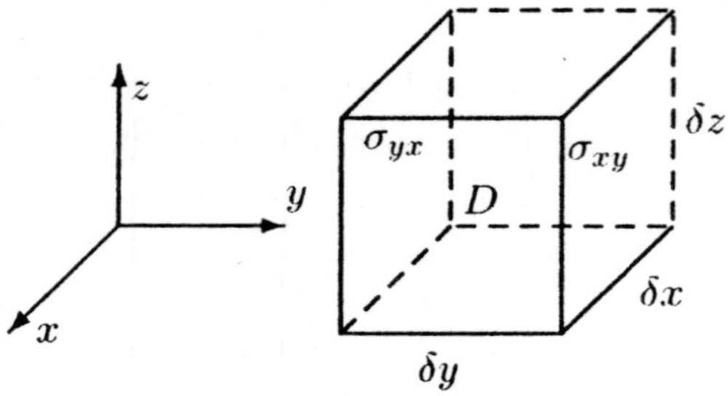


Figure 13.1: A cubic element in an elastostatic body.

Using the stress-strain relationship, it is straightforward to prove that the strain tensor is also symmetric ($\varepsilon_{ij} = \varepsilon_{ji}$). In engineering, this tensor is often written as

$$\boldsymbol{\sigma} = \begin{pmatrix} \sigma_x & \tau_{xy} & \tau_{xz} \\ \tau_{xy} & \sigma_y & \tau_{yz} \\ \tau_{xz} & \tau_{yz} & \sigma_z \end{pmatrix}, \tag{13.33}$$

so as to emphasize that the non-diagonal elements are for shear components $\tau_{xy}$ etc.

From the matrix algebra, we know that a square matrix can always be expressed in terms of eigenvalues and eigenvectors. For the stress tensor, the eigenvalue problem

$$(\boldsymbol{\sigma} - \sigma_i I)\tilde{\mathbf{n}} = 0, \tag{13.34}$$

provides the the principal stresses $\sigma_i (i = 1, 2, 3)$ (eigenvalues) and their principal directions $\tilde{\mathbf{n}}^T = \mathbf{n_1}, \mathbf{n_2}, \mathbf{n_3}$ (eigenvectors). In the coordinate system formed by the three eigenvectors, the stress is expressed by the three principal stresses along three principal directions, and there are no shear stress components.

The non-trivial solutions require the determinant of the coefficient matrix must be zero. That is

$$\begin{vmatrix} \sigma_x - \sigma & \tau_{xy} & \tau_{xz} \\ \tau_{xy} & \sigma_y - \sigma & \tau_{yz} \\ \tau_{xz} & \tau_{yz} & \sigma_z - \sigma \end{vmatrix} = 0. \tag{13.35}$$

It can be expanded into a cubic equation

$$\sigma^3 - (\sigma_x + \sigma_y + \sigma_z)\sigma^2$$

$$+[\sigma_x\sigma_y + \sigma_y\sigma_z + \sigma_z\sigma_x - (\tau_{xy}^2 + \tau_{yz}^2 + \tau_{xz}^2)]\sigma$$

$$-[\sigma_x\sigma_y\sigma_z + 2\tau_{xy}\tau_{yz}\tau_{xz} - (\sigma_x\tau_{yz}^2 + \sigma_y\tau_{xz}^2 + \sigma_z\tau_{xy}^2)] = 0. \tag{13.36}$$

There are three invariants ($I_1$, $I_2$ and $I_3$) for the second-order symmetric stress tensor $\sigma_{ij}$, and these invariants satisfy the characteristic equation

$$\sigma^3 - I_1\sigma^2 + I_2\sigma - I_3 = 0, \tag{13.37}$$

where

$$I_1 = \mathrm{tr}(\sigma_{ij}) = \sigma_x + \sigma_y + \sigma_z, \tag{13.38}$$

$$I_2 = \sigma_x\sigma_y + \sigma_y\sigma_z + \sigma_z\sigma_x - (\tau_{xy}^2 + \tau_{yz}^2 + \tau_{xz}^2), \tag{13.39}$$

and

$$I_3 = \sigma_x\sigma_y\sigma_z + 2\tau_{xy}\tau_{yz}\tau_{xz} - (\sigma_x\tau_{yz}^2 + \sigma_y\tau_{xz}^2 + \sigma_z\tau_{xy}^2). \tag{13.40}$$

Under appropriate transformations, this tensor can be transformed into a diagonal form

$$\sigma_{ij} \mapsto \begin{pmatrix} \sigma_1 & 0 & 0 \\ 0 & \sigma_2 & 0 \\ 0 & 0 & \sigma_3 \end{pmatrix}, \tag{13.41}$$

where $\sigma_1, \sigma_2, \sigma_3$ are principal stresses. Written in terms of principal stresses, the three invariants become

$$I_1 = \sigma_1 + \sigma_2 + \sigma_3, \tag{13.42}$$

$$I_2 = \sigma_1\sigma_2 + \sigma_2\sigma_3 + \sigma_3\sigma_1, \tag{13.43}$$

and

$$I_3 = \sigma_1\sigma_2\sigma_3. \tag{13.44}$$

Now consider an elastic body when $n$ concentrated load $\mathbf{f_1}, \mathbf{f_2}, ..., \mathbf{f_n}$ acted upon the body at $n$ different points. The displacements at each point in the direction of the corresponding force are $\mathbf{q}_1, \mathbf{q}_2, ..., \mathbf{q}_n$. For a linear elastic body, the principle of superposition applies and we have

$$\mathbf{q}_1 = C_{11}\boldsymbol{f}_1 + C_{12}\boldsymbol{f}_2 + ... + C_{1n}\boldsymbol{f}_n, \tag{13.45}$$

$$\mathbf{q}_2 = C_{21}\boldsymbol{f}_1 + C_{22}\boldsymbol{f}_2 + ... + C_{2n}\boldsymbol{f}_n, \tag{13.46}$$

$$...$$

$$\mathbf{q}_n = C_{n1}\boldsymbol{f}_1 + C_{n2}\boldsymbol{f}_2 + ... + C_{nn}\boldsymbol{f}_n, \tag{13.47}$$

where $C_{ij}$ are the influence coefficients or flexibility matrix. The total work done due to this set of loads is

$$W = \frac{1}{2}\sum_{i=1}^{n} \mathbf{f_i} \cdot \mathbf{q_i}. \tag{13.48}$$

There is a very useful theorem concerning these coefficients. It is called Maxwell's reciprocal theorem or Maxwell-Betti theorem, which states that the influence coefficients (or flexibility) matrix is symmetric $C_{ij} = C_{ji}$. That is to say, the displacement at point $i$ due to a unit load at another point $j$ is equal to the displacement at $j$ due to a unit load at point $i$. This theorem is essentially equivalent to say the displacements are path-independent and independent of the order of the loads applied upon the elastostatic body.

For two forces $\boldsymbol{f}_i$ and $\boldsymbol{f}_j$, the final displacements are the same where $\boldsymbol{f}_i$ is applied first, then $\boldsymbol{f}_j$, or $\boldsymbol{f}_j$ is applied first then $\boldsymbol{f}_i$, or even both are applied at the same time. In other words, the system has no memory of the load history. In the case of only two forces, we first apply $\boldsymbol{f}_i$ slowly (so as to reduce the dynamical effect) with $\boldsymbol{f}_j = 0$, the displacement of point $i$ is $C_{ii}\boldsymbol{f}_i$ and the displacement of point $j$ is $C_{ji}\boldsymbol{f}_i$. The work done is $\frac{1}{2}C_{ii}f_i^2$ where $f_i^2 = \mathbf{f_i} \cdot \mathbf{f_i}$. Now with $\boldsymbol{f}_i$ kept fixed, we apply $\boldsymbol{f}_j$ slowly, the additional displacement at point $i$ is $C_{ij}\boldsymbol{f}_j$ and the additional displacement at point $j$ is $C_{jj}\boldsymbol{f}_j$. In this case, the extra work done is $C_{ij}\mathbf{f_i} \cdot \mathbf{f_j} + \frac{1}{2}C_{jj}f_j^2$. The total work done at the final state is

$$W = \frac{1}{2}C_{ii}f_i^2 + \frac{1}{2}C_{jj}f_j^2 + C_{ij}(\mathbf{f_i} \cdot \mathbf{f_j}). \tag{13.49}$$

If follow the same procedure but slowly apply $\boldsymbol{f}_j$ first, then $\boldsymbol{f}_i$, the total work done is now

$$\bar{W} = \frac{1}{2}C_{ii}f_i^2 + \frac{1}{2}C_{jj}f_j^2 + C_{ji}(\mathbf{f_i} \cdot \mathbf{f_j}). \tag{13.50}$$

As the total work done should be independent of the order in which the loads are applied, this requires $W = \bar{W}$, which leads to

$$C_{ij}(\mathbf{f_i} \cdot \mathbf{f_j}) = C_{ji}(\mathbf{f_i} \cdot \mathbf{f_j}). \tag{13.51}$$

Since $\mathbf{f_i} \cdot \mathbf{f_j}$ is a dot product and thus a scalar, we now get

$$C_{ij} = C_{ji}. \tag{13.52}$$

This completes the proof of the Maxwell's reciprocal theorem. This theorem is the important basis for boundary element analysis and virtual work method in computational engineering.

## 13.3 Equations of Motion

For a general solid where the inertia is not negligible, we have

$$\nabla \cdot \boldsymbol{\sigma} + \mathbf{b} = \rho \frac{\partial^2 \mathbf{u}}{\partial t^2}, \tag{13.53}$$

where $\rho$ is the density of the elastic body. In some books, the following form of body force $\mathbf{b} = \rho \mathbf{f}$ is used, in this case, the force $\boldsymbol{f}$ means the force per unit mass. Together with the generalized Hooke's law and relationship with displacement $\mathbf{u}$, we have the following set of equations of motion for an elastic body.

$$\frac{\partial \sigma_{ij}}{\partial x_j} + b_i = \rho \frac{\partial^2 u_i}{\partial t^2}. \quad (13.54)$$

$$\sigma_{ij} = 2G\varepsilon_{ij} + \lambda \varepsilon_{kk}\delta_{ij}, \quad (13.55)$$

$$\varepsilon_{ij} = \frac{1}{2}(\frac{\partial u_i}{\partial x_j} + \frac{\partial u_j}{\partial x_i}). \quad (13.56)$$

There are 15 equations (6 for stresses, 6 for strains, 3 for displacement) and we have 15 unknowns (6 stress components, 6 strain components and 3 displacements). Therefore, the elastic field should be uniquely determined if appropriate boundary conditions are given. There are other compatibility equations as well, and we will briefly discuss them later.

If we write the equations of motion using the bold font notations, we have

$$\nabla \cdot \boldsymbol{\sigma} + \mathbf{b} = \rho \frac{\partial^2 \boldsymbol{u}}{\partial t^2},, \quad (13.57)$$

$$\boldsymbol{\sigma} = 2G\boldsymbol{\varepsilon} + \lambda(\nabla \cdot \boldsymbol{u})\boldsymbol{\delta}, \quad (13.58)$$

$$\boldsymbol{\varepsilon} = \frac{1}{2}(\nabla \boldsymbol{u} + \nabla \boldsymbol{u}^T). \quad (13.59)$$

If we substitute the generalized Hooke's law and displacement into the first equation (13.57), we have

$$\nabla \cdot [2G\boldsymbol{\varepsilon} + \lambda(\nabla \cdot \boldsymbol{u})\boldsymbol{\delta}] + \mathbf{b} = \rho \frac{\partial^2 \boldsymbol{u}}{\partial t^2}, \quad (13.60)$$

or

$$\nabla \cdot [G(\nabla \boldsymbol{u} + \nabla \boldsymbol{u}^T) + \lambda(\nabla \cdot \boldsymbol{u})\boldsymbol{\delta}] + \mathbf{b} = \rho \frac{\partial^2 \boldsymbol{u}}{\partial t^2}, \quad (13.61)$$

which leads to

$$(G + \lambda)\nabla(\nabla \cdot \boldsymbol{u}) + G\nabla^2 \boldsymbol{u} + \mathbf{b} = \rho \frac{\partial^2 \boldsymbol{u}}{\partial t^2}. \quad (13.62)$$

Using $G + \lambda = G/(1 - 2\nu)$ and after some rearrangements, we have

$$\rho \frac{\partial^2 \boldsymbol{u}}{\partial t^2} = \frac{G}{1 - 2\nu} \nabla(\nabla . \boldsymbol{u}) + G\nabla^2 \boldsymbol{u} + \mathbf{b}, \tag{13.63}$$

which is the well-known Cauchy-Navier equation. This equation supports both longitudinal wave ($P$ wave) and transverse wave ($S$ wave). In the simplest 1-D case without any (external) body force $\mathbf{b}$, we can take $\nabla \cdot \boldsymbol{u} = 0$ for $S$-wave, the equation is simplified as

$$\rho \frac{\partial^2 u_1}{\partial t^2} = G \frac{\partial^2 u_1}{\partial x^2}, \tag{13.64}$$

thus its wave speed is

$$v_S = \sqrt{\frac{G}{\rho}}. \tag{13.65}$$

For the $P$-wave in 1-D, the displacement field is non-rotational, *i.e.*, $\nabla \times (\nabla \times \boldsymbol{u}) = 0$. From the identity $\nabla(\nabla \cdot \boldsymbol{u}) = \nabla \times (\nabla \times \boldsymbol{u}) + \nabla^2 \boldsymbol{u}$, the 1-D Cauchy-Navier equation becomes

$$\rho \frac{\partial^2 u_1}{\partial t^2} = (\lambda + 2G) \frac{\partial^2 u_1}{\partial x^2}. \tag{13.66}$$

Then, the speed of $P$-wave is

$$v_P = \sqrt{\frac{(\lambda + 2G)}{\rho}}. \tag{13.67}$$

Since $\lambda + 2G > G$, therefore, $P$-waves always travel faster than $S$-waves.

Furthermore, from the definitions of the strain components in terms of displacements $\boldsymbol{u}^T = (u_1, u_2, u_3) = (u, v, w)$, we have

$$\varepsilon_{xx} = \frac{\partial u}{\partial x}, \qquad \varepsilon_{yy} = \frac{\partial v}{\partial y}, \qquad \varepsilon_{xy} = \frac{1}{2}(\frac{\partial u}{\partial y} + \frac{\partial v}{\partial x}). \tag{13.68}$$

By assuming the displacements are continuous and differentiable functions of positions, we differentiate $\varepsilon_{xx}$ with respect to $y$ twice, we have

$$\frac{\partial^2 \varepsilon_{xx}}{\partial y^2} = \frac{\partial^3 u}{\partial x \partial y^2}. \tag{13.69}$$

Similarly, differentiate $\varepsilon_{yy}$ with respect to $x$ twice, we have

$$\frac{\partial^2 \varepsilon_{yy}}{\partial x^2} = \frac{\partial^3 v}{\partial y \partial x^2}. \tag{13.70}$$

Now differentiate $\varepsilon_{xy}$ with respect to $y$ once, and with respect to $x$ once, we have

$$\frac{\partial \varepsilon_{xy}}{\partial x \partial y} = \frac{1}{2}[\frac{\partial^3 u}{\partial x \partial y^2} + \frac{\partial^3 v}{\partial y \partial x^2}] = \frac{1}{2}[\frac{\partial^2 \varepsilon_{xx}}{\partial y^2} + \frac{\partial^2 \varepsilon_{yy}}{\partial x^2}], \tag{13.71}$$

where we have used the interchangeability of partial derivatives $\partial^2 v/\partial x \partial y = \partial^2 v/\partial y \partial x$. This can be rearranged as

$$\frac{\partial^2 \varepsilon_{xx}}{\partial y^2} + \frac{\partial^2 \varepsilon_{yy}}{\partial x^2} = 2\frac{\partial^2 \varepsilon_{xy}}{\partial x \partial y}, \tag{13.72}$$

which is the compatibility equation. In the same fashion, we can derive other compatibility equations

$$\frac{\partial^2 \varepsilon_{zz}}{\partial y^2} + \frac{\partial^2 \varepsilon_{yy}}{\partial z^2} = 2\frac{\partial^2 \varepsilon_{yz}}{\partial y \partial z}. \tag{13.73}$$

$$\frac{\partial^2 \varepsilon_{xx}}{\partial z^2} + \frac{\partial^2 \varepsilon_{zz}}{\partial x^2} = 2\frac{\partial^2 \varepsilon_{xz}}{\partial x \partial z}. \tag{13.74}$$

## 13.4 Airy Stress Functions

For certain engineering problems, the solutions in a plane is of concern. In this case, we are dealing with plane strain and plane stress problems. For a plane stress problem, we assume that $\sigma_{zz} = 0$ (but $\varepsilon_{zz} \neq 0$), then the plane stress problem involves no stress components depending on $z$. That is to say $\sigma_{xz} = \sigma_{yz} = \sigma_{zz} = 0$. We have only three independent stress components $\sigma_{xx}, \sigma_{yy}$, and $\sigma_{xy}$. The generalized Hooke's law reduces to

$$\varepsilon_{xx} = \frac{1}{E}(\sigma_{xx} - \nu\sigma_{yy}), \quad (13.75)$$

$$\varepsilon_{yy} = \frac{1}{E}(\sigma_{yy} - \nu\sigma_{xx}), \quad (13.76)$$

$$\varepsilon_{xy} = \frac{1+\nu}{E}\sigma_{xy}. \quad (13.77)$$

However,

$$\varepsilon_{zz} = \frac{-\nu}{1-\nu}(\varepsilon_{xx} + \varepsilon_{yy}), \quad (13.78)$$

which is not zero in general.

For plane strain problems, it is assumed that $\varepsilon_{zz} = 0$. Thus, there are only three independent strain components $\varepsilon_{xx}, \varepsilon_{yy}$, and $\varepsilon_{xy}$, however, the stress $\sigma_{zz} = \nu(\sigma_{xx} + \sigma_{yy})$ is not zero. The compatibility equation becomes

$$\frac{\partial^2 \varepsilon_{xx}}{\partial y^2} + \frac{\partial^2 \varepsilon_{yy}}{\partial x^2} = 2\frac{\partial^2 \varepsilon_{xy}}{\partial x \partial y}. \quad (13.79)$$

For plane strain problems with no body forces, the equilibrium equations are automatically satisfied if the stress components are related to a scalar function $\Phi$, called Airy's stress function. The Airy's stress function is defined by

$$\sigma_{xx} = \frac{\partial^2 \Phi}{\partial y^2}, \qquad \sigma_{yy} = \frac{\partial^2 \Phi}{\partial x^2}, \qquad \sigma_{xy} = -\frac{\partial^2 \Phi}{\partial x \partial y}. \quad (13.80)$$

In this case, the compatibility equation becomes

$$\nabla^2(\nabla^2\Phi) = 0, \quad (13.81)$$

which is a biharmonic equation and can be written as

$$\nabla^4\Phi = 0. \quad (13.82)$$

In cylindrical polar coordinates $(r, \theta, z)$, it becomes

$$[\frac{\partial^2}{\partial r^2} + \frac{1}{r}\frac{\partial}{\partial r} + \frac{1}{r^2}\frac{\partial^2}{\partial \theta^2}]^2\Phi = 0. \quad (13.83)$$

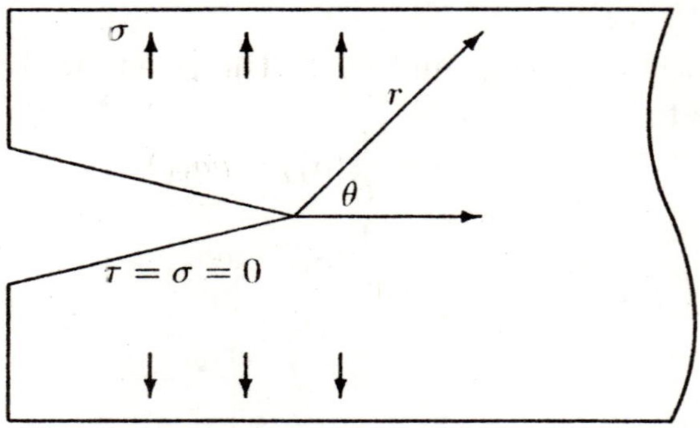


Figure 13.2: A crack in an elastic plate.

Now consider a semi-infinite crack in an infinite solid as shown in Figure 13.2, assuming the elastic body deforms in plane strain. The far field is subjected to bounded stress at infinity. The surfaces of the crack shall be stress free, which leads to the following boundary conditions

$$\sigma_{\theta\theta} = \frac{\partial^2 \Phi}{\partial r^2} = 0, \ \sigma_{r\theta} = -\frac{\partial}{\partial r}(\frac{1}{r}\frac{\partial \Phi}{\partial \theta}) = 0, \text{ at } \quad \theta = \pm\pi. \quad (13.84)$$

Let us try a solution of the form

$$\Phi = r^{n+1} f(\theta), \quad (13.85)$$

and substitute it into the governing biharmonic equation, we get

$$[\frac{d^2}{d\theta^2} + (n+1)^2][\frac{d^2}{d\theta^2} + (n-1)^2]f(\theta) = 0. \quad (13.86)$$

As the second-order equation $y'' + \lambda^2 y = 0$ has a general solution $y = A\sin\lambda\theta + B\cos\lambda\theta$, we can here use this method twice, the general solution takes the following form

$$f(\theta) = A\cos(n+1)\theta + B\sin(n+1)\theta$$

$$+C\cos(n-1)\theta + D\sin(n-1)\theta. \quad (13.87)$$

The boundary conditions become

$$\sigma_{\theta\theta} = r^{n-1} n(n+1) f(\theta), \quad (13.88)$$

and

$$\sigma_{r\theta} = r^{n-1}n\{(n+1)[A\sin(n+1)\theta - B\cos(n+1)\theta]$$

$$+(n-1)[C\sin(n-1)\theta - D\cos(n-1)\theta]\}, \tag{13.89}$$

at $\theta = \pm\pi$. We know $n = 0$ is trivial. From the first equation, we have

$$\sin(2n\pi) = 0, \qquad n = \pm\frac{1}{2}, \pm 1, \pm\frac{3}{2}, ..., \tag{13.90}$$

and $r^n (n \geq 1)$ does not converge, therefore, they are not suitable solutions. The constraint now becomes $n \leq 0$, but the solutions has singularity as $r \to 0$. This is however acceptable in the crack propagation as the stress concentrations do physically exist. Substituting the general solution into the boundary conditions with $n = 1/2$ and $\theta = \pm\pi$, we get

$$3A + C = 0, \qquad B - D = 0. \tag{13.91}$$

By defining the stress intensity factor $K_I$ for the crack,

$$K_I = \frac{3A\sqrt{2\pi}}{4}, \tag{13.92}$$

which is for the opening (model I) of the crack. It is a limit of stress at $\theta = 0$

$$K_I = \lim_{r \to 0} \sigma_{\theta\theta}(r, \theta)\Big|_{\theta=0}. \tag{13.93}$$

Finally, the solution of stresses can be written as

$$\sigma_{rr} = \frac{K_I}{\sqrt{2\pi r}}(1 + \sin^2\frac{\theta}{2})\cos\frac{\theta}{2}, \tag{13.94}$$

$$\sigma_{\theta\theta} = \frac{K_I}{\sqrt{2\pi r}}\cos^3\frac{\theta}{2}, \tag{13.95}$$

$$\sigma_{r\theta} = \frac{K_I}{\sqrt{2\pi r}}\cos^2\frac{\theta}{2}\sin\frac{\theta}{2}. \tag{13.96}$$

Once we have the stress distribution, we can get the strains. Then the displacements are the integration of the strains, and we have

$$u_r = \frac{K_I(1-\nu)}{E}\sqrt{\frac{2r}{\pi}}[(5-4\nu)\cos\frac{\theta}{2} - \cos\frac{3\theta}{2}], \tag{13.97}$$

and

$$u_\theta = \frac{K_I(1-\nu)}{E}\sqrt{\frac{2r}{\pi}}[\sin\frac{3\theta}{2} - (5-4\nu)\sin\frac{\theta}{2}]. \tag{13.98}$$

## 13.5 Euler-Bernoulli Beam Theory

The Euler-Bernoulli beam theory is a simplified theory for calculating the deflection of beams under a distribution of load force using the linear isotropic theory of elasticity. The basic assumptions for the beam theory are: 1) the beam is isotropic and elastic; 2) the beam deformation is dominated by bending, and distortion and rotation are negligible; 3) the beam is long and slender with a constant cross section along the axis. Under these assumptions, we can now derive the governing equations.

Let $u(x,t)$ be the deflection of the beam (shown in Figure 13.3), $A$ be the area of the cross section, and $f(x,t)$ be the force per unit length. The first assumption implies that the bending moment $M$ is proportional to the curvature $\kappa$ of the bending. That is

$$M = EI\kappa, \qquad \kappa = \frac{\frac{\partial^2 u}{\partial x^2}}{[1+(\frac{\partial^2 u}{\partial x^2})^2]^{3/2}}, \tag{13.99}$$

where $E$ is the Young's modulus and $I$ is the area moment of the beam's cross section. In mechanics, $I$ is also called the second moment of area or the area moment of inertia. It is worth pointing out that the area moment about a horizontal axis through the centroid is defined by

$$I = \int_\Omega y^2 dA, \tag{13.100}$$

which has a unit of $[\text{m}]^4$, and it should not be confused with the mass moment of inertia $J$ (also often denoted as $I$, but we use $J$ here) about an axis, which is defined by

$$J = \int_\Omega r^2 dm = \int_\Omega \rho r^2 dxdydz \tag{13.101}$$

with a unit of [Kg] $[\text{m}]^2$. Both $E$ and $I$ do not change along the $x$-axis. For a cylindrical rod with a radius of $R$, we have $I = \pi R^4/4$. For a rectangular beam with a base width of $b$ and a depth of $h$, we have $I = bh^3/12$.

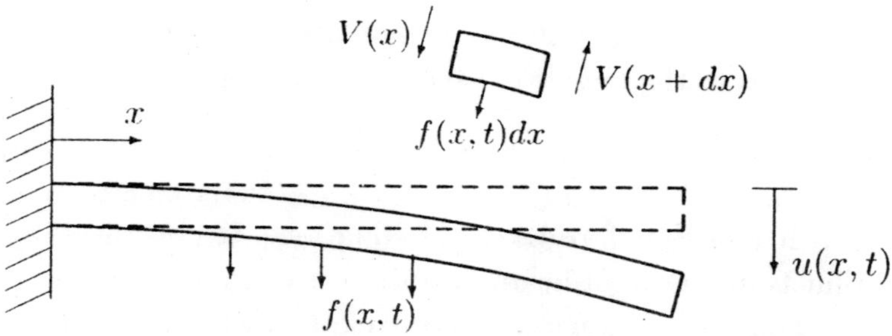


Figure 13.3: Beam bending.

The second assumption means that the shear $V(x)$ is related to the bending moment

$$\frac{\partial M}{\partial x} = V(x), \tag{13.102}$$

and the third assumption means $\frac{\partial u}{\partial x} \ll 1$. Therefore, we have

$$M \approx EI \frac{\partial^2 u}{\partial x^2}, \tag{13.103}$$

or

$$V \approx \frac{\partial}{\partial x}(EI \frac{\partial^2 u}{\partial x^2}). \tag{13.104}$$

For a small volume element (also shown in Figure 13.3), the mass of the element is $\rho A dx$ where $\rho$ is the density, and the acceleration is $\frac{\partial^2 u}{\partial t^2}$. The shear force variation $V(x + dx) = V(x) + \frac{\partial V}{\partial x} dx$, and the total force is

$$V(x) - V(x + dx) + f(x,t)dx = [f(x,t) - \frac{\partial V}{\partial x}]dx. \quad (13.105)$$

Using the Newton's second law of motion, we have

$$f(x,t) - \frac{\partial V}{\partial x} = \rho A \frac{\partial^2 u}{\partial t^2}. \quad (13.106)$$

Substituting the above expression for $V$, we have

$$\rho A \frac{\partial^2 u}{\partial t^2} + \frac{\partial^2}{\partial x^2}[EI \frac{\partial^2 u}{\partial x^2}] = f(x), \quad (13.107)$$

which is the Euler-Bernoulli equation. If there is no force $f(x,t) = 0$, the equation becomes a homogeneous form

$$\rho A \frac{\partial^2 u}{\partial t^2} + \frac{\partial^2}{\partial x^2}[EI \frac{\partial^2 u}{\partial x^2}] = 0, \quad (13.108)$$

which is a fourth-order wave equation. It governs the waves that travel along a beam, a rod or any slender column.

This equation can essentially explain why spaghetti and dry pasta almost always break into more than two fragments. You can try in your kitchen to break a slender spaghetti by holding its two ends and gradually form an arc and bend beyond its curvature limit. When a spaghetti rod snaps, it will generally break three to pieces. This phenomenon is very interesting and once puzzled the famous physicist Richard Feynman for quite a while, and it was recently studied by two scientists that the brittle fragmentation process is virtually governed by this Cauchy-Navier equation. They found that the sudden relaxation of the curvature by first breaking will lead to a burst of flexural waves along the spaghetti rod, and these waves locally increase the curvature in the rod, resulting in more fragmented pieces.

For the elastostatic problem, $\frac{\partial^2 u}{\partial t^2} \approx 0$, we have

$$\frac{\partial^2}{\partial x^2}[EI \frac{\partial^2 u}{\partial x^2}] = q(x), \quad (13.109)$$

where $q(x) = f(x)$ is the applied force per unit length. This equation will be used to determine the deflection of a beam.

---

□ ***Example 13.1:*** *Let us now use the Euler-Bernoulli theory to calculate the shape of a heavy cantilever with a uniform cross section under its own gravity. For a beam under its own gravity, the force is constant $q(x) = \rho g A$ per unit length where $g$ is the acceleration due to the Earth's gravity. If the length of the cantilever is $L$, then the total weight is $W = \rho g A L$, thus $q = \frac{W}{L}$. Therefore, we have*

$$EI\frac{d^4u}{dx^4} = q = \frac{W}{L}$$

*where we have use $EI = const$. Integrating it twice, we have*

$$EI\frac{d^2u}{dx^2} = \frac{q}{2}x^2 + Ax + B.$$

*At the free end, the beam cannot support bending moment and/or shear, which implies that $M(L) = 0$ and $V(L) = q * L + A = 0$. These conditions lead to $A = -qL$ and $B = qL^2/2$. Integrating the above equation again, we have*

$$EI\frac{du}{dx} = \frac{q}{6}x^3 - \frac{qL}{2}x^2 + \frac{qL^2}{2}x + C.$$

*As the beam is fixed at $x = 0$, we have $u = 0$ and $\frac{du}{dx} = 0$ at $x = 0$. Thus we have $C = 0$ from $u = 0$. Integrating once again, we have*

$$EIu = \frac{q}{24}x^4 - \frac{qL}{6}x^3 + \frac{qL^2}{4}x^2 + D.$$

*As $\acute{u}_x = 0$ at $x = 0$, we have $D = 0$. Therefore, the final deflection curve becomes*

$$u = \frac{1}{EI}[\frac{q}{24}x^4 - \frac{qL}{6}x^3 + \frac{qL}{4}x^2].$$

*The end deflection $\delta = u(x = L)$ is*

$$\delta = \frac{qL^4}{8EI} = WL^3/(8EI).$$

□

---

# Chapter 14

# Mathematical Models

What we have discussed so far in terms of differential equations is very limited. In engineering and natural sciences, there are so many different kinds of phenomena that require both mathematical modelling and computer simulations as well as experimental studies. In most cases, the classical models (using the heat conduction equation and the wave equation and others) are simply not adequate to described these phenomena. Therefore, we have to broaden our view to study other kinds of partial differential equations. In fact, mathematical modelling per se is a subject with vast literature, and subsequently we have to focus on the relevant equations and to introduce them very briefly.

## 14.1 Classic Models

Before we introduce more complicated partial differential equations, let us first remind us the three types of classic partial differential equations because they are widely used and occur in a vast range of applications. To a certain extent, almost all books or studies on the partial differential equations will have to deal with these three types of basic partial differential equations.

### 14.1.1 Laplace's and Poisson's Equation

In heat transfer problems, the steady state of heat conduction with a source is governed by the Poison equation

$$k\nabla^2 u = f(x, y, t), \qquad (x, y) \in \Omega, \tag{14.1}$$

or

$$u_{xx} + u_{yy} = q(x, y, t), \tag{14.2}$$

for two independent variables $x$ and $y$. Here $k$ is thermal diffusivity and $f(x, y, t)$ is the heat source. If there is no heat source ($q = 0$), this becomes the Laplace equation. The solution or a function is said to be harmonic if it satisfies Laplace's equation.

In order to determine the temperature $u$ completely, the appropriate boundary conditions are needed. A simple boundary condition is to specify the temperature $u = u_0$ on the boundary $\partial\Omega$. This type of problem is the Dirichlet problem. On the other hand, if the temperature is not known, but the gradient $\partial u/\partial n$ is known on the boundary where $\mathbf{n}$ is the outward-pointing unit normal, and this forms the Neumann problem. Furthermore, some problems may have a mixed type of boundary conditions in the combination of $\alpha u + \beta\frac{\partial u}{\partial n} = \gamma$, which naturally occur as a radiation or cooling boundary condition.

### 14.1.2 Parabolic Equation

Time-dependent problems, such as diffusion and transient heat conduction, are governed by the parabolic equation

$$u_t = k u_{xx}. \tag{14.3}$$

Written in the $n$-dimensional case $x_1 = x, x_2 = y, x_3 = z, ...,$ it can be extended to the reaction-diffusion equation

$$u_t = k\nabla^2 u + f(x_1, .., x_n, t), \tag{14.4}$$

where $f$ is the reaction rate.

### 14.1.3 Wave Equation

The vibration of strings and travelling sound waves are governed by the hyperbolic wave equation. The 1-D wave equation in its simplest form is

$$u_{tt} = c^2 u_{xx}, \tag{14.5}$$

where $c$ is the velocity of the wave. Using a transformation of the pair of independent variables

$$\xi = x + ct, \qquad \eta = x - ct, \tag{14.6}$$

for $t > 0$ and $-\infty < x < \infty$, the wave equation can be written as

$$u_{\xi\eta} = 0. \tag{14.7}$$

Integrating twice and substituting back in terms of $x$ and $t$, we have

$$u(x, t) = f(x + ct) + g(x - ct), \tag{14.8}$$

where $f$ and $g$ are arbitrary functions of $x + ct$ and $x - ct$, respectively. We can see that there are two directions that the wave can travel. One wave moves to the right and one travels to the left at a constant speed $c$.

## 14.2 Other PDEs

We have shown examples of the three major equations of second-order linear partial differential equations. There are other equations that occur frequently in mathematical physics, engineering and computational sciences. We will give a brief description of some of these equations.

### 14.2.1 Elastic Wave Equation

As we have seen in the linear elasticity, the wave in an elastic isotropic homogeneous solid is governed by the following equation in terms of displacement $\mathbf{u}$,

$$\rho \frac{\partial^2 \mathbf{u}}{\partial t^2} = \mu \nabla^2 \mathbf{u} + (\lambda + \mu) \nabla(\nabla \cdot \mathbf{u}) + \mathbf{b}, \tag{14.9}$$

where $\rho$ is density. $\lambda$ and $\mu$ are Lamé constants. $\mathbf{b}$ is body force. Such an equation can have two types of wave: transverse wave (S-wave) and longitudinal or dilatational wave (P-wave).

### 14.2.2 Maxwell's Equations

The scientific essence of modern wireless communications is governed by the Maxwell's equations for electromagnetic waves

$$\nabla \cdot \mathbf{E} = \frac{\rho_e}{\varepsilon_0}, \tag{14.10}$$

$$\nabla \times \mathbf{E} = -\frac{\partial \mathbf{B}}{\partial t}, \tag{14.11}$$

$$\nabla \cdot \mathbf{B} = 0, \tag{14.12}$$

$$\nabla \times \mathbf{B} = \mu_0 \mathbf{J} + \frac{1}{c^2} \frac{\partial \mathbf{E}}{\partial t}, \tag{14.13}$$

where $\rho_e$ is the charge density, $\mathbf{E}$ is the electric field, $\mathbf{B}$ is the magnetic field and $\mathbf{J}$ is the current density. $\varepsilon_0$ and $\mu_0$ are the permittivity and permeability of the free space, respectively. Finally, $c$ is the speed of light. The first equation is the electrostatic equation, the second one is the Faraday's law and the last equation is the Ampere-Maxwell's law.

### 14.2.3 Reaction-Diffusion Equation

The reaction-diffusion equation is an extension of heat conduction with a source

$$u_t = D\Delta u + f(x, y, z, u), \tag{14.14}$$

where $D$ is the diffusion coefficient and $f$ is the reaction rate. One example is the combustion equation

$$u_t = Du_{xx} + Que^{-\lambda/u}, \tag{14.15}$$

where $D, Q$ and $\lambda$ are constants. The other example is the Fitz-Hugh-Nagumo equations for transport of a nerve signal

$$u_t = u_{xx} + u(1-u)(u-\alpha) - v, \tag{14.16}$$

$$v_t = \epsilon(u - \gamma v), \tag{14.17}$$

where $\gamma > 0$, $0 < \alpha < 1$ and $\epsilon \ll 1$. These equations are sometimes also called the equations in excitable media.

### 14.2.4 Fokker-Plank Equation

The time evolution of the probability density function of position and velocity of a particle system is described by the Fokker-Plank equation

$$\frac{\partial p}{\partial t} = [-\sum_{i=1}^{n} \frac{\partial D_i^{[1]}}{\partial x_i} + \sum_{i=1}^{n}\sum_{j=1}^{n} \frac{\partial^2 D_{ij}^{[2]}}{\partial x_i \partial x_j}]p, \tag{14.18}$$

which can be generalized as

$$\frac{\partial p}{\partial t} = \sum_{k=1}^{N} \sum_{i_1,...,i_k} (-1)^k \frac{\partial}{\partial x_{i_1}} ... \frac{\partial}{\partial x_{i_k}} \{D_{i_1,...,i_k}^{[k]} p\}, \tag{14.19}$$

where $D^{[k]}(x_1, x_2, ..., x_n)$ are tensors. In the special case of $N = 2$, $D^{[1]}(x_1, ..., x_n)$ is the drift vector, while $D^{[2]}$ is the diffusion tensor.

### 14.2.5 Black-Scholes Equation

In the option pricing model, the value of an option $u(S, t)$ at time $t$ is governed by the well-known Nobel-winning Black-Scholes equation

$$\frac{\partial u}{\partial t} = ru - rS\frac{\partial u}{\partial S} - \frac{1}{2}\sigma^2 S^2 \frac{\partial^2 u}{\partial S^2}, \tag{14.20}$$

where $S$ is the current stock price of the underlying stock and $r$ is the risk-free interest rate. $t$ is the time until option expiration. $\sigma$ is the stock volatility or the standard deviation of stock returns. The Black-Scholes equation is very similar to the extended version of the diffusion equation. The interesting feature for a call option with an exercise price $E$ and expiry time $T$, the change of variables

$$\tau = (T-t)\sigma^2/2, \qquad v = ue^{r(T-t)} \tag{14.21}$$

and

$$x = \ln(S/E) + (\frac{2r}{\sigma^2} - 1)\tau, \tag{14.22}$$

can transform it into a standard diffusion equation

$$\frac{\partial v}{\partial \tau} = \frac{\partial^2 v}{\partial x^2}. \tag{14.23}$$

Then, we can use the standard methods such as integral transforms to solve this equation.

### 14.2.6 Schrödinger Equation

The famous Schrödinger equation is the revolutionary equation in quantum mechanics and molecular dynamics

$$i\hbar\frac{\partial \Psi}{\partial t} = -\frac{\hbar^2}{2m}\nabla^2\Psi + U\Psi, \tag{14.24}$$

where $\hbar$ is a Planck constant. This equation can be obtained from the energy form $E = \frac{p^2}{2m} + U$ (where $p$ is the momentum) using differential operator mapping $E \to i\hbar\frac{\partial}{\partial t}$ and $p \to -i\hbar\nabla$. $\Psi$ is the probability wave function and $U = V(r) - E$ is the energy potential. This is a complex partial differential equation.

### 14.2.7 Navier-Stokes Equations

The Navier-Stokes equations for an incompressible flow can be written as

$$\nabla \cdot \mathbf{u} = 0, \tag{14.25}$$

$$\mathbf{u}_t + (\mathbf{u} \cdot \nabla)\mathbf{u} = \frac{1}{\mathrm{Re}}\nabla^2\mathbf{u} - \nabla p, \tag{14.26}$$

where $\mathrm{Re} = \rho \mathrm{UL}/\mu$ is the Reynolds number. $U$ is the typical velocity and $L$ is the length scale. $\rho$ and $\mu$ are the density of the fluid and its viscosity, respectively. In computational fluid dynamics, most simulations are mainly related to these equations.

In the limit of $\mathrm{Re} \ll 1$, we have the Stokes flow (slow flow) governed by

$$\mu \nabla^2 \mathbf{u} = \nabla p. \tag{14.27}$$

In another limit $\mathrm{Re} \gg 1$, we have the inviscous flow

$$\nabla \cdot \mathbf{u} = 0, \qquad \mathbf{u}_t + (\mathbf{u} \cdot \nabla)\mathbf{u} = -\nabla p. \tag{14.28}$$

We can see that the equations are still nonlinear even in this simplified case.

### 14.2.8 Sine-Gordon Equation

Another important equation that appears in a wide range of applications in physics and many other fields is the Sine-Gordon equation

$$u_{tt} - u_{xx} + \sin(u) = 0, \tag{14.29}$$

which can generally be written as

$$u_{tt} = u_{xx} + \alpha \sin(\omega u). \tag{14.30}$$

This is a nonlinear hyperbolic equation.

Almost all these equations are very difficult for mathematical analysis, and most of them do not have closed-form solutions under most common boundary conditions. Therefore, the numerical methods are a good alternative in this case. In the next few chapters, we will introduce various numerical methods in details.

# Chapter 15

# Finite Difference Method

The finite difference method is one of the most popular methods that are used commonly in computer simulations. It has the advantage of simplicity and clarity, especially in 1-D configuration and other cases with regular geometry. The finite difference method essentially transforms an ordinary differential equation into a set of algebraic equations by replacing the continuous derivatives with finite difference approximations on a grid of mesh or node points that spans the domain of interest based on the Taylor expansions. In general, the boundary conditions and boundary nodes need special treatment.

## 15.1 Integration of ODEs

The second-order or higher order ordinary differential equations can be written as a first-order system of ODEs. Since the technique for solving a system is essentially the same as that for solving a single equation

$$\frac{dy}{dx} = f(x, y), \tag{15.1}$$

then we shall focus on the first-order equation in the rest of this section. In principle, the solution can be obtained by direct integration,

$$y(x) = y_0 + \int_{x_0}^{x} f(x, y(x))dx, \tag{15.2}$$

but in practice it is usually impossible to do the integration analytically as it requires the solution of $y(x)$ to evaluate the right-hand side. Thus, some approximations shall be utilized. Numerical integration is the most common technique to obtain approximate solutions. There are various integration schemes with different orders of accuracy and convergent rates. These schemes include the simple Euler scheme, Runge-Kutta method, Relaxation method, and many others.

### 15.1.1 Euler Scheme

Using the notations $h = \Delta x = x_{n+1} - x_n$, $y_n = y(x_n)$, $x_n = x_0 + n\Delta x$ $(n = 0, 1, 2, ..., N)$, and $' = d/dx$ for convenience, then the explicit Euler scheme can simply be written as

$$y_{n+1} = y_n + \int_{x_n}^{x_{n+1}} f(x, y)dx \approx y_n + hf(x_n, y_n). \tag{15.3}$$

This is a forward difference method as it is equivalent to the approximation of the first derivative

$$y'_n = \frac{y_{n+1} - y_n}{\Delta x}. \tag{15.4}$$

The order of accuracy can be estimated using the Taylor expansion

$$y_{n+1} = y_n + hy'|_n + \frac{h^2}{2}y''|_n + ...$$

$$\approx y_n + hf(x_n, y_n) + O(h^2). \tag{15.5}$$

Thus, the Euler method is first order accurate.

For any numerical algorithms, the algorithm must be stable in order to reach convergent solutions. Thus, stability is an important issue in numerical analysis. Defining $\delta y$ as the discrepancy between the actual numerical solution and the true solution of the Euler finite difference equation, we have

$$\delta y_{n+1} = [1 + hf'(y)] = \xi \delta y_n. \tag{15.6}$$

In order to avoid the discrepancy to grow, it requires the following stability condition $|\xi| \leq 1$. The stability restricts the size of interval $h$, which is usually small. One alternative that can use larger $h$ is the implicit Euler scheme, and this scheme approximates the derivative by a backward difference $y'_n = (y_n - y_{n-1})/h$ and the right-hand side of equation (15.2) is evaluated at the new $y_{n+1}$ location. Now the scheme can be written as

$$y_{n+1} = y_n + hf(x_{n+1}, y_{n+1}). \tag{15.7}$$

The stability condition becomes

$$\delta y_{n+1} = \xi \delta y_n = \frac{\delta y_n}{1 - hf'(y)}, \tag{15.8}$$

which is always stable if $f'(y) = \frac{\partial f}{\partial y} \leq 0$. This means that any step size is acceptable. However, the step size cannot be too large as the accuracy reduces as the step size increases. Another practical issue is that, for most problems such as nonlinear ODEs, the evaluation of $y'$ and $f'(y)$ requires the value of $y_{n+1}$ which is unknown. Thus, an iteration procedure is needed to march to a new value $y_{n+1}$, and the iteration starts with a guess value which is usually taken to be zero for most cases. The implicit scheme generally gives better stability.

---

□ ***Example 15.1:*** *To solve the equation*

$$\frac{dy}{dx} = f(y) = e^{-y} - y,$$

*we use the explicit Euler scheme, and we have*

$$y_{n+1} \approx y_n + hf(y_n) = y_n + h(e^{-y_n} - y_n).$$

*Suppose the discrepancy between real solution $y_n^*$ and the numerical $y_n$ is $\delta y_n$ so that $y_n^* = y_n + \delta y_n$, then the real solution satisfies*

$$y_{n+1}^* = y_n^* + hf(y_n^*).$$

*Since $f(y_n^*) = f(y_n) + \frac{df}{dy}\delta y_n$, the above equation becomes*

$$y_{n+1} + \delta y_{n+1} = y_n + \delta y_n + h[f(y_n) + f'(y_n)\delta y_n].$$

*Together with the Euler scheme, we have*

$$\delta y_{n+1} = \delta y_n + f'\delta y_n.$$

*Suppose that $\delta y_n \propto \xi^n$, then we have*

$$\xi^{n+1} = \xi^n + hf'\xi^n, \qquad \text{or} \qquad \xi = 1 + hf'.$$

*In order for the scheme to be stable (or $\xi^n \to 0$), it requires that*

$$|\xi| \le 1, \qquad \text{or} \qquad -1 \le 1 + hf' = 1 - h(e^{-y_n} + 1) \le 1.$$

*The stability condition becomes*

$$0 \le h \le \frac{2}{e^{-y_n} + 1}.$$

□

### 15.1.2 Leap-Frog Method

The Leap-frog scheme is the central difference

$$y_n' = \frac{y_{n+1} - y_{n-1}}{2\Delta x}, \tag{15.9}$$

which leads to

$$y_{n+1} = y_{n-1} + 2hf(x_n, y_n). \tag{15.10}$$

The central difference method is second order accurate. In a similar way as equation (15.6), the leap frog method becomes

$$\delta y_{n+1} = \delta y_{n-1} + 2hf'(y)\delta y_n, \tag{15.11}$$

or

$$\delta y_{n+1} = \xi^2 \delta y_{n-1}, \tag{15.12}$$

where $\xi^2 = 1 + 2hf'(y)\xi$. This scheme is stable only if $|\xi| \le 1$, and a special case is $|\xi| = 1$ when $f'(y)$ is purely imaginary. Therefore, the central scheme is not necessarily a better scheme than the forward scheme.

### 15.1.3 Runge-Kutta Method

We have so far seen that stability of the Euler method and the central difference method is limited. The Runge-Kutta method uses a trial step to the midpoint of the interval by central difference and combines with the forward difference at two steps

$$\hat{y}_{n+1/2} = y_n + \frac{h}{2} f(x_n, y_n), \tag{15.13}$$

$$y_{n+1} = y_n + h f(x_{n+1/2}, \hat{y}_{n+1/2}). \tag{15.14}$$

This scheme is second order accurate with higher stability compared with previous simple schemes. One can view this scheme as a predictor-corrector method. In fact, we can use multisteps to devise higher order methods if the right combinations are used to eliminate the error terms order by order. The popular classical Runge-Kutta method can be written as

$$a = h f(x_n, y_n),$$

$$b = h f(x_n + h/2, y_n + a/2),$$

$$c = h f(x_n + h, y_n + b/2),$$

$$d = h f(x_n + h, y_n + c),$$

$$y_{n+1} = y_n + \frac{a + 2(b + c) + d}{6}, \tag{15.15}$$

which is fourth order accurate. Generally speaking, the higher-order scheme is better than the lower scheme, but not always.

## 15.2 Hyperbolic Equations

Numerical solutions of partial differential equations are more complicated than that of ODEs because it involves time and space variables and the geometry of the domain of interest. Usually, boundary conditions are more complex. In addition, nonlinear problems are very common in engineering applications. Now we start with the simplest first order equations and then move onto more complicated cases.

### 15.2.1 First-Order Hyperbolic Equation

For simplicity, we start with the one-dimensional scalar equation of hyperbolic type,

$$\frac{\partial u}{\partial t} + c\frac{\partial u}{\partial x} = 0, \tag{15.16}$$

where $c$ is a constant or the velocity of advection. By using the forward Euler scheme for time and centered-spaced scheme, we have

$$\frac{u_j^{n+1} - u_j^n}{\Delta t} + c[\frac{u_{j+1}^n - u_{j-1}^n}{2h}] = 0, \tag{15.17}$$

where $t = n\Delta t, n = 0, 1, 2, ..., x = x_0 + jh, j = 0, 1, 2, ...,$ and $h = \Delta x$. In order to see how this method behaves numerically, we use the von Neumann stability analysis.

Assuming the independent solutions or eigenmodes (also called Fourier modes) in spatial coordinate $x$ in the form of $u_j^n = \xi^n e^{ikhj}$, and substituting into equation (15.17), we have

$$\xi^{n+1}e^{ikhj} - \xi^n e^{ikhj} = \xi^n \frac{c\Delta t}{h}\frac{e^{ikh(j+1)} - e^{ihk(j-1)}}{2}. \tag{15.18}$$

Dividing both sides of the above equation by $\xi^n \exp(ikhj)$ and using $\sin x = (e^{ix} - e^{-ix})/2i$, we get

$$\xi = 1 - i\frac{c\Delta t}{h}\sin(kh). \tag{15.19}$$

The stability criteria $|\xi| \leq 1$ require

$$(\frac{c\Delta t}{h})^2 \sin^2 kh \leq 0. \tag{15.20}$$

However, this inequality is impossible to satisfy and this scheme is thus unconditionally unstable.

To avoid the difficulty of instability, we can use other schemes such as the upwind scheme and Lax scheme. For the upwind scheme, the equation becomes

$$\frac{u_j^{n+1} - u_j^n}{\Delta t} + c[\frac{u_j^n - u_{j-1}^n}{h}] = 0, \tag{15.21}$$

whose stability condition is

$$|\xi| = |1 - \frac{c\Delta t}{h}[1 - \cos(kh) + i\sin(kh)]| \leq 1, \tag{15.22}$$

which is equivalent to

$$0 < \frac{c\Delta t}{h} \leq 1. \tag{15.23}$$

This the well-known Courant-Friedrichs-Lewy stability condition, often referred to as the Courant stability condition. Thus, the upwind scheme is conditionally stable.

### 15.2.2 Second-Order Wave Equation

Higher order equations such as the second-order wave equation can be written as a system of hyperbolic equations and then be solved using numerical integration. They can also be solved by direct discretization using a finite difference scheme. The wave equation

$$\frac{\partial^2 u}{\partial t^2} = c^2 \frac{\partial^2 u}{\partial x^2}, \tag{15.24}$$

consists of second derivatives. If we approximate the first derivatives at each time step $n$ using

$$u_i' = \frac{u_{i+1}^n - u_i^n}{\Delta x}, \quad u_{i-1}' = \frac{u_i^n - u_{i-1}^n}{\Delta x}, \tag{15.25}$$

then we can use the following approximation for the second derivative

$$u_i'' = \frac{u_i' - u_{i-1}'}{\Delta x}$$

$$= \frac{u_{i+1}^n - 2u_i^n + u_{i-1}^n}{(\Delta x)^2}. \tag{15.26}$$

This is in fact a central difference scheme of second order accuracy. If we use the similar scheme for time-stepping, then we get a central difference scheme in both time and space.

Thus, the numerical scheme for this equation becomes

$$\frac{u_i^{n+1} - 2u_i^n + u_i^{n-1}}{(\Delta t)^2} = c^2 \frac{u_{i+1}^n - 2u_i^n + u_{i-1}^n}{(\Delta x)^2}. \tag{15.27}$$

This is a two-level scheme with a second order accuracy. The idea of sol· ing this difference equation is to express (or to solve) $u_i^{n+1}$ at time step $t = n+1$ in terms of the known values or data $u_i^n$ and $u_i^{n-1}$ at two previous time steps $t = n$ and $t = n - 1$.

## 15.3 Parabolic Equation

For the parabolic equation such as the diffusion or heat conduction equation

$$\frac{\partial u}{\partial t} = \frac{\partial}{\partial x}(D\frac{\partial u}{\partial x}), \tag{15.28}$$

a simple Euler method for the time derivative and centered second-order approximations for space derivatives lead to

$$u_j^{n+1} = u_j^n + \frac{D\Delta t}{h^2}(u_{j+1}^n - 2u_j^n + u_{j-1}^n). \tag{15.29}$$

The stability requirement $\xi \leq 1$ leads to the constraint on the time step (see the example),

$$\Delta t \leq \frac{h^2}{2D}. \tag{15.30}$$

This scheme is shown in Figure 15.1 and it is conditionally stable.

---

□ ***Example 15.2:*** *From equation (15.29), we can apply the von Neumann stability analysis by assuming $u_j^n = \xi^n e^{ikhj}$, we have*

$$\xi^{n+1} e^{ikhj} = \xi^n e^{ikhj} + \frac{D\Delta x}{h^2}\xi^n [e^{ikh(j+1)} - 2e^{ikhj} + e^{ikh(j-1)}].$$

*Dividing both sides by $\xi^n e^{ikhj}$, we have*

$$\xi = 1 + \frac{D\Delta t}{h^2}[e^{ikh} + e^{-ikh} - 2].$$

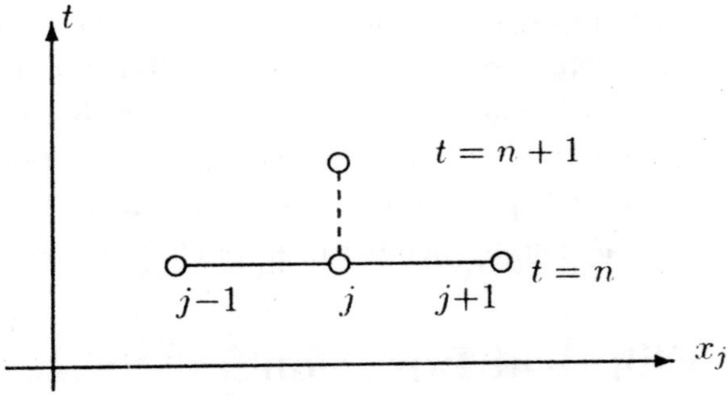


Figure 15.1: Central difference in space and explicit Euler time-stepping.

*Using* $\cos x = (e^{ix} + e^{-ix})/2$ *and* $\sin^2(x/2) = (1 - \cos x)/2$, *we obtain*

$$\xi = 1 - \frac{4D\Delta t}{h^2}\sin^2(\frac{kh}{2}).$$

*Since* $\sin(x) \leq 1$, *thus* $\xi \leq 1$ *requires*

$$-1 \leq 1 - \frac{4D\Delta t}{h^2} \leq 1,$$

*or*

$$0 \leq \Delta t \leq \frac{h^2}{2D}.$$

□

---

A typical feature of a solution to the diffusive system is that the profile is gradually smoothed out as time increases. The time-stepping scheme we used limits the step size of time as larger time steps will make the scheme unstable. There are many ways to improve this, and one of most widely used schemes is the implicit scheme.

To avoid the difficulty caused by very small timesteps, we now use an implicit scheme for time derivative differencing, and thus we have

$$u_j^{n+1} - u_j^n = \frac{D\Delta t}{h^2}(u_{j+1}^{n+1} + 2u_j^{n+1} + u_{j-1}^{n+1}). \tag{15.31}$$

Applying the stability analysis, we have

$$\xi = \frac{1}{1 + \frac{4D\Delta t}{h^2}\sin^2\frac{kh}{2}}, \tag{15.32}$$

whose norm is always less than unity ($|\xi| \leq 1$). This means the implicit scheme is unconditionally stable for any size of time steps. That is why implicit methods are more desired in simulations. However, there is one disadvantage of this method, which requires more programming skills because the inverse of a large matrix is usually needed in implicit schemes.

## 15.4 Elliptical Equation

In the parabolic equation, if the time derivative is zero or $u$ does not change with time $u_t = 0$, then it becomes a steady-state problem that is governed by the elliptic equation. For the steady state heat conduction problem, we generally have the Poisson problem,

$$\nabla \cdot [\kappa(u, x, y, t)\nabla u] = f, \tag{15.33}$$

If $\kappa$ is a constant, this becomes

$$\nabla^2 u = q, \qquad q = \frac{f}{\kappa}. \tag{15.34}$$

There are many methods available to solve this problems such as the boundary integral method, the relaxation method, and the multigrid method. Two major ones are the long-time approximation of the transient parabolic diffusion equations, the other includes the iteration method.

The long time approximation method is essentially based on fact that the parabolic equation

$$\frac{\partial u}{\partial t} + \kappa\nabla^2 u = f, \tag{15.35}$$

evolves with a typical scale of $\sqrt{\kappa t}$. If $\sqrt{\kappa t} \gg 1$, the system is approaching its steady state. Assuming $t \to \infty$ and $\kappa \gg 1$, we then have

$$\nabla^2 u = \frac{f}{\kappa} - \frac{1}{\kappa}u_t \to 0. \tag{15.36}$$

The long-time approximation is based on the fact that the parabolic equation in the case of $\kappa$ = const degenerates into the above steady-state equation (15.33) because $u_t \to 0$ as $t \to \infty$. This approximation becomes better if $\kappa \gg 1$. Thus, the usual numerical methods for solving parabolic equations are valid. However, other methods may obtain the results more quickly.

The iteration method uses the second-order scheme for space derivatives, and equation (15.34) in the 2-D case becomes

$$\frac{u_{i+1,j} - 2u_{i,j} + u_{i-1,j}}{(\Delta x)^2} + \frac{u_{i,j+1} - 2u_{i,j} + u_{i,j-1}}{(\Delta y)^2} = q. \quad (15.37)$$

If we use $\Delta x = \Delta y = h$, then the above equation simply becomes

$$(u_{i,j+1} + u_{i,j-1} + u_{i+1,j} + u_{i-1,j}) - 4u_{i,j} = h^2 q, \quad (15.38)$$

which can be written as

$$\mathbf{A}\mathbf{u} = \mathbf{b}. \quad (15.39)$$

In principle, one can solve this equation using Gauss elimination; however, this becomes impractical as the matrix becomes large such as $1000 \times 1000$. The Gauss-Seidel iteration provides a more efficient way to solve this equation by splitting $\mathbf{A}$ as

$$\mathbf{A} = \mathbf{L} + \mathbf{D} + \mathbf{U}, \quad (15.40)$$

where $\mathbf{L}, \mathbf{D}, \mathbf{U}$ are the lower triangle, diagonal and upper triangle matrices of $\mathbf{A}$, respectively. The iteration is updated in the following way:

$$\mathbf{u}^{(n)} = (\mathbf{D} + \mathbf{L})^{-1}[\mathbf{b} - \mathbf{U}\mathbf{u}^{(n-1)}]. \quad (15.41)$$

This procedure stops until a prescribed error or precision is reached.

# Chapter 16

# Finite Volume Method

## 16.1 Introduction

The finite difference method discussed in the previous chapter approximates the ordinary differential equations and partial differential equations using Taylor series, resulting in a system of algebraic equations. The finite volume method resembles the finite difference method in certain ways but the starting point is the integral formulation of the problem. It uses the integral form of the partial differential equations in terms of conservation laws, then approximates the surface and boundary integrals in the control volumes. This becomes convenient for problems involving flow or flux boundaries.

For a hyperbolic equation that is valid in the domain $\Omega$ with boundary $\partial\Omega$,

$$\frac{\partial u}{\partial t} - \nabla \cdot (\kappa \nabla u) = q, \tag{16.1}$$

or written in terms of flux function $\mathbf{F} = \mathbf{F}(u) = -\kappa \nabla u$, we have

$$\frac{\partial u}{\partial t} + \nabla \cdot \mathbf{F} = q. \tag{16.2}$$

The integral form of this equation becomes

$$\int_\Omega \frac{\partial u}{\partial t} d\Omega + \int_\Omega \nabla \cdot \mathbf{F} = \int_\Omega q d\Omega. \tag{16.3}$$

If the integral form is decomposed into many small control volumes, or finite volumes, $\Omega = \bigcup_{i=1}^{N} \Omega_i$ and $\Omega_i \bigcap \Omega_j = \emptyset$. By defining the control volume cell average or mean value

$$u_i = \frac{1}{V_i}\int_{\Omega_i} u d\Omega_i, \qquad q_i = \frac{1}{V_i}\int_{\Omega_i} q d\Omega_i, \tag{16.4}$$

where $V_i = |\Omega_i|$ is the volume of the small control volume $\Omega_i$, the above equation can be written as

$$\frac{\partial u_i}{\partial t} + \sum_{i=1}^{N} \frac{1}{V_i}\int_{\Omega_i} \nabla \cdot \mathbf{F}(u_i) d\Omega_i = q_i, \tag{16.5}$$

By using the divergence theorem

$$\int_V \nabla \cdot \mathbf{F} = \int_\Gamma \mathbf{F} \cdot \mathbf{n} dA, \tag{16.6}$$

we have

$$\frac{\partial u_i}{\partial t} + \sum_{i=1}^{N} \frac{1}{V_i}\int_{\Gamma_i} \mathbf{F} \cdot \mathbf{dS} = q_i, \tag{16.7}$$

where $\mathbf{dS} = \mathbf{n} dA$ is the surface element and $\mathbf{n}$ is the outward pointing unit vector on the surface $\Gamma_i$ enclosing the finite volume $\Omega_i$. The integration can be approximated using various numerical integration schemes. In the simplest 1-D case with $h = \Delta x$, the integration

$$u_i = \frac{1}{h}\int_{(i-1/2)h}^{(i+1/2)h} u dx, \tag{16.8}$$

is a vertex-centred finite volume scheme. In the following sections, we will discuss the three major types of partial differential equations (elliptic, parabolic and hyperbolic) and their finite volume discretizations.

## 16.2 Elliptic Equations

Laplace's equation is one of the most studied elliptic equations

$$\nabla^2 u(x, y) = 0, \qquad (x, y) \in \Omega, \tag{16.9}$$

its integral form becomes

$$\int_\Omega \nabla^2 u d\Omega = \int_\Gamma \frac{\partial u}{\partial \mathbf{n}} \cdot \mathbf{dS} = 0. \tag{16.10}$$

For the simple regular grid points $(i\Delta x, j\Delta y)$, the control volume in this case is a cell centred at $(i\Delta x, j\Delta y)$ with a size of $\Delta x$ (along $x$-axis) and $\Delta y$ (along $y$-axis), and the boundary integral on any cell consists of four parts integrated on each of the four sides. By using the simple approximation $\frac{\partial u}{\partial n}$ with $\frac{\partial u}{\partial x} = (u_{i+1,j} - u_{i,j})/\Delta x$ and $\frac{\partial u}{\partial y} = (u_{i,j+1} - u_{i,j})/\Delta y$, we have

$$\int_{\Omega_{i,j}} \frac{\partial u}{\partial n} d\Omega = \frac{\Delta y}{\Delta x}(u_{i+1,j} + u_{i-1,j} - 2u_{i,j})$$

$$+ \frac{\Delta x}{\Delta y}(u_{i,j+1} + u_{i,j-1} - 2u_{i,j}) = 0. \tag{16.11}$$

Dividing both sides with $\Delta x \Delta y$, and letting $\Delta x = \Delta y = h$, we obtain

$$(u_{i+1,j} + u_{i,j+1} + u_{i-1,j} + u_{i,j-1}) - 4u_{i,j} = 0, \tag{16.12}$$

which resembles finite difference methods in many ways. In fact, this is exactly the Laplace operator for a 5-point differencing scheme.

## 16.3 Parabolic Equations

For the case of heat conduction

$$\frac{\partial u}{\partial t} = k \frac{\partial^2 u}{\partial x^2} + q(u, x, t), \tag{16.13}$$

we have its integral form

$$\int_t \int_\Omega \left(\frac{\partial u}{\partial t} - k \frac{\partial^2 u}{\partial x^2} - q\right) dx dt = 0. \tag{16.14}$$

If we use the control volume from $(i-1/2)h$ to $(i+1/2)h$ where $h = \Delta x$, and with time from step $n$ to $n+1$, we have

$$\int_{n\Delta t}^{(n+1)\Delta t} \int_{(i-1/2)h}^{(i+1/2)h} (\frac{\partial u}{\partial t} - k\frac{\partial^2 u}{\partial x^2} - q)dxdt = 0. \tag{16.15}$$

By using the mid-point approximation

$$\int_a^b \psi(x)dx = \psi[\frac{(a+b)}{2}](b-a), \tag{16.16}$$

and the DuFort-Frankel scheme where we first approximate the gradient

$$\frac{\partial^2 u}{\partial x^2} = \frac{u_{i+1}^n - 2u_i^n + u_{i-1}^n}{h^2}, \tag{16.17}$$

then replace $-2u_i^n$ with $-(u_j^{n+1} + u_j^{n-1})$, we have

$$\frac{u_i^{n+1} - u_i^{n-1}}{2\Delta t} =$$

$$\frac{[(u_{i+1}^n - (u_i^{n+1} + u_i^{n-1}) + u_{i-1}^n)]}{h^2} + q_i^n, \tag{16.18}$$

where we have used the central scheme for time as well. The finite volume scheme is more versatile in dealing with irregular geometry and more natural in applying boundary conditions. Following the stability analysis, we get $|\xi| < 1$ is always true and thus the Dufort-Frankel scheme is unconditionally stable for all $\Delta t$ and $\Delta x$.

## 16.4 Hyperbolic Equations

For the hyperbolic equation of the conservation law in the one-dimensional case

$$\frac{\partial u}{\partial t} + \frac{\partial \Psi(u)}{\partial x} = 0, \tag{16.19}$$

we have its integral form in the fixed domain

$$\int_{x_a}^{x_b} \frac{\partial u}{\partial t} dx = \frac{\partial}{\partial t} \int_{x_a}^{x_b} u dx$$

$$= -\{\Psi[u(x_b)] - \Psi[u(x_a)]\} = 0. \tag{16.20}$$

If we use the mid-point rule $u^*$ to approximate the integral, we have

$$(x_b - x_a)\frac{\partial u^*}{\partial t} = -\{\Psi[u(x_b)] - \Psi[u(x_a)]\}. \tag{16.21}$$

If we choose the control volume $[(i-1/2)\Delta x, (i+1/2)\Delta x]$ centred at the mesh point $x_i = i\Delta x = ih$ with the approximation $u_i \approx u_i^*$ in each interval, and using the forward differencing scheme for the time derivative, we have

$$u_i^{n+1} - u_i^n = -\frac{\Delta t}{h}[\Psi(x_{i+1/2}) - \Psi(x_{i-1/2})]. \tag{16.22}$$

By further approximation of the flux $\Psi(x_{i+1/2}) \approx \Psi(x_i)$, we have the upward scheme

$$u_i^{n+1} - u_i^n = -\frac{\Delta t}{h}[\Psi(u_i) - \Psi(u_{i-1})], \tag{16.23}$$

which is conditionally stable as we know this in the finite difference method. For the simplest flux $\Psi(u) = cu$, we have

$$u^{n+1} = u_i^n - \frac{c\Delta t}{h}(u_i^n - u_{i-1}^n), \tag{16.24}$$

and its stability requires that

$$0 < \frac{c\Delta t}{h} \leq 1. \tag{16.25}$$

# Chapter 17

# Finite Element Method

In the finite difference method, we approximate the equations at a finite number of discrete points, and there are many limitations in finite difference methods. One of such disadvantages is that it is not straightforward to deal with irregular geometry. More versatile and efficient methods are highly needed. In fact, the finite element method is one class of the most successful methods in engineering and have a wide range of applications.

The basic aim of the finite element method is to formulate the numerical method in such a way that the partial differential equation will be transformed into algebraic equations in terms of matrices. For time-dependent problems involving partial differential equations, the equations can be converted into an ordinary differential equation, which will in turn be discretized and converted into algebraic equations by time-stepping or some iteration techniques. For example, a linear elastic problem can be formulated in such a way that it is equivalent to the equation of the following type

$$\mathbf{K}\mathbf{u} = \mathbf{f}. \tag{17.1}$$

where $\mathbf{K}$ is the stiffness matrix, and $\mathbf{f}$ is a vector corresponding to nodal forces and some contribution from boundary conditions. $\mathbf{u}$ is the unknown vector to be solved and it corresponds to the nodal degree of freedom such as the displacement.

# 17.1 Concept of Elements

## 17.1.1 Simple Spring Systems

The basic idea of the finite element analysis is to divide a model (such as a bridge and an airplane) into many pieces or elements with discrete nodes. These elements form an approximate system to the whole structures in the domain of interest, so that the physical quantities such as displacements can be evaluated at these discrete nodes. Other quantities such as stresses, strains can then be evaluated at certain points (usually Gaussian integration points) inside elements. The simplest elements are the element with two nodes in 1-D, the triangular element with three nodes in 2-D, and tetrahedral elements with four nodes in 3-D.

In order to show the basic concept, we now focus on the simplest 1-D spring element with two nodes (see Figure 17.1). The spring has a stiffness constant $k$ (N/m) with two nodes $i$ and $j$. At nodes $i$ and $j$, the displacements (in metres) are $u_i$ and $u_j$, respectively. $f_i$ and $f_j$ are nodal forces.

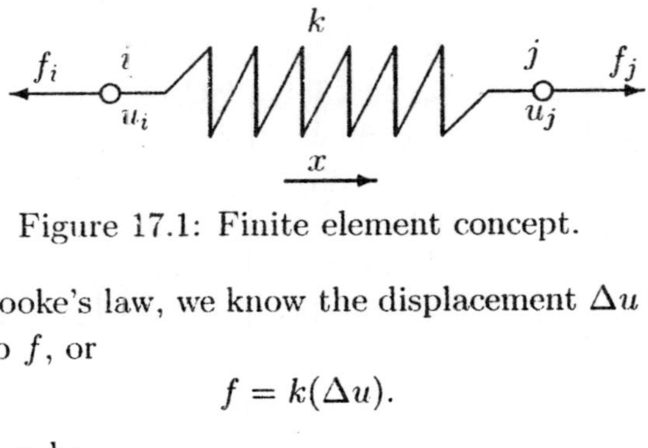


Figure 17.1: Finite element concept.

From Hooke's law, we know the displacement $\Delta u = u_j - u_i$ is related to $f$, or

$$f = k(\Delta u). \tag{17.2}$$

At node $i$, we have

$$f_i = -f = -k(u_j - u_i) = ku_i - ku_j, \tag{17.3}$$

and at node $j$, we get

$$f_j = f = k(u_j - u_i) = -ku_i + ku_j. \tag{17.4}$$

These two equations can be combined into a matrix equation

$$\begin{pmatrix} k & -k \\ -k & k \end{pmatrix} \begin{pmatrix} u_i \\ u_j \end{pmatrix} = \begin{pmatrix} f_i \\ f_j \end{pmatrix}, \qquad \text{or} \qquad \mathbf{K}\mathbf{u} = \mathbf{f}. \quad (17.5)$$

Here $\mathbf{K}$ is the stiffness matrix, $\mathbf{u}$ and $\mathbf{f}$ are the displacement vector and force vector, respectively. This is the basic spring element, and let us see how it works in a spring system such as shown in Figure 17.2 where three different springs are connected in series.

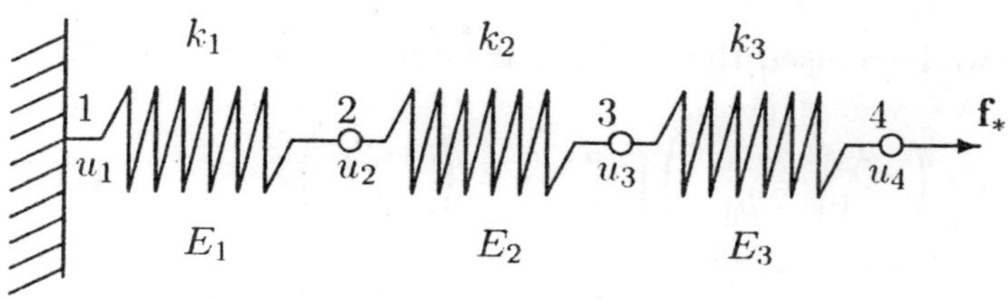


Figure 17.2: A simple spring system.

For a simple spring system shown in Figure 17.2, we now try to determine the displacements of $u_i (i = 1, 2, 3, 4)$. In order to do so, we have to assemble the whole system into a single equation in terms of global stiffness matrix $\mathbf{K}$ and forcing $\boldsymbol{f}$. As these three elements are connected in series, the assembly of the system can be done element by element. For element $E_1$, its contribution to the overall global matrix is

$$\begin{pmatrix} k_1 & -k_1 \\ -k_1 & k_1 \end{pmatrix} \begin{pmatrix} u_1 \\ u_2 \end{pmatrix} = \begin{pmatrix} f_1 \\ f_2 \end{pmatrix}, \qquad (17.6)$$

which is equivalent to

$$\mathbf{K}_1 \mathbf{u} = \mathbf{f}_{E_1}, \qquad (17.7)$$

where

$$\begin{pmatrix} k_1 & -k_1 & 0 & 0 \\ -k_1 & k_1 & 0 & 0 \\ 0 & 0 & 0 & 0 \\ 0 & 0 & 0 & 0 \end{pmatrix} \begin{pmatrix} u_1 \\ u_2 \\ u_3 \\ u_4 \end{pmatrix} = \begin{pmatrix} f_1 \\ f_2 \\ . \\ . \end{pmatrix}, \qquad (17.8)$$

and $\mathbf{f}_{E_1}^T = (f_1, f_2, 0, 0)$. Similarly, for element $E_2$, we have

$$\begin{pmatrix} k_2 & -k_2 \\ -k_2 & k_2 \end{pmatrix} \begin{pmatrix} u_2 \\ u_3 \end{pmatrix} = \begin{pmatrix} -f_2 \\ f_3 \end{pmatrix}, \tag{17.9}$$

or

$$\mathbf{K_2} = \begin{pmatrix} 0 & 0 & 0 & 0 \\ 0 & k_2 & -k_2 & 0 \\ 0 & -k_2 & k_2 & 0 \\ 0 & 0 & 0 & 0 \end{pmatrix}, \tag{17.10}$$

where we have used the balance at node 2. For element $E_3$, we have

$$\begin{pmatrix} k_3 & -k_3 \\ -k_3 & k_3 \end{pmatrix} \begin{pmatrix} u_3 \\ u_4 \end{pmatrix} = \begin{pmatrix} -f_3 \\ f_* \end{pmatrix}, \tag{17.11}$$

or

$$\mathbf{K}_3 = \begin{pmatrix} 0 & 0 & 0 & 0 \\ 0 & 0 & 0 & 0 \\ 0 & 0 & k_3 & -k_3 \\ 0 & 0 & -k_3 & k_3 \end{pmatrix}, \tag{17.12}$$

where $f_4 = f_*$ has been used. We can now add the three sets of equations together to obtain a single equation

$$\begin{pmatrix} k_1 & -k_2 & 0 & 0 \\ -k_1 & k_1 + k_2 & -k_2 & 0 \\ 0 & -k_2 & k_2 + k_3 & -k_3 \\ 0 & 0 & -k_3 & k_3 \end{pmatrix} \begin{pmatrix} u_1 \\ u_2 \\ u_3 \\ u_4 \end{pmatrix} = \begin{pmatrix} f_1 \\ -f_2 + f_2 \\ -f_3 + f_3 \\ f_* \end{pmatrix},$$

or

$$\mathbf{Ku} = \mathbf{f}, \tag{17.13}$$

where

$$\mathbf{K} = \mathbf{K}_1 + \mathbf{K}_2 + \mathbf{K}_3$$

$$= \begin{pmatrix} k_1 & -k_1 & 0 & 0 \\ -k_1 & k_1 + k_2 & -k_2 & 0 \\ 0 & -k_2 & k_2 + k_3 & -k_3 \\ 0 & 0 & -k_3 & k_3 \end{pmatrix}, \tag{17.14}$$

and

$$\mathbf{u}^T = (u_1, u_2, u_3, u_4), \qquad \mathbf{f} = \mathbf{f}_{E_1} + \mathbf{f}_{E_2} + \mathbf{f}_{E_3}. \tag{17.15}$$

In general, the matrix $\mathbf{K}$ is singular or its rank is less than the total number of degrees of freedom, which is four in this case. This means that the equation has no unique solution. Thus, we need the boundary conditions to ensure a unique solution. In this spring system, if no boundary condition is applied at any nodes, then the applied force at the node 4 will make the spring system fly to the right. If we add a constraint by fixing the left node 1, then the system can stretch, and a unique configuration is formed.

In our case where there are no applied forces at nodes 2 and 3, we have

$$\mathbf{f}^T = (0, 0, 0, f_*). \tag{17.16}$$

---

□ ***Example 17.1:*** *For $k_1 = 100$ N/m, $k_2 = 200$ N/m, and $k_3 = 50$N/m, and $f_* = 20$ N, the boundary at node 1 is fixed ($u_1 = 0$). Then, the stiffness matrix is*

$$\mathbf{K} = \begin{pmatrix} 100 & -100 & 0 & 0 \\ -100 & 300 & -200 & 0 \\ 0 & -200 & 250 & -50 \\ 0 & 0 & -50 & 50 \end{pmatrix},$$

*and the force column vector*

$$\mathbf{f}^T = (0, 0, 0, 20).$$

*The rank of $\mathbf{K}$ is 3, therefore, we need at least one boundary condition. By applying $u_1 = 0$, we now have only three unknown displacements $u_2, u_3, u_4$. Since $u_1 = 0$ is already known, the first equation for $u_1$ becomes redundant and we can now delete it so that the reduced stiffness matrix $\mathbf{A}$ is a $3 \times 3$ matrix. Therefore, we have*

$$\mathbf{A} = \begin{pmatrix} 300 & -200 & 0 \\ -200 & 250 & 0 \\ 0 & -50 & 50 \end{pmatrix},$$

*and the reduced forcing vector is*

$$\mathbf{g}^T = (0, 0, 20).$$

*The solution is*

$$\mathbf{u} = \mathbf{A}^{-1}\mathbf{g} = \begin{pmatrix} 0.2 \\ 0.3 \\ 0.7 \end{pmatrix}.$$

*Therefore, the displacements are $u_2 = 0.2$ m, $u_3 = 0.3$ m, and $u_4 = 0.7$ m.*

*Theoretically speaking, the force should be 20N everywhere in the spring systems since the mass of the springs is negligible. Let us calculate the force at nodes 2 and 3 to see if this is the case. At the node 2, the extension in element $E_1$ is $\Delta u = u_2 - u_1 = 0.2$ m, thus the force at node 2 is*

$$f_2 = k_1 \Delta u = 100 \times 0.2 = 20\text{N}.$$

*Similarly, at node 3 of element $E_2$, we have*

$$f_3 = k_2(u_3 - u_2) = 200 \times 0.1 = 20\text{N},$$

*which is the same at node 3 of element $E_3$*

$$f_3 = k_3 \times (-\Delta u) = k_3(u_4 - u_3) = 50 \times 0.4 = 20\text{N}.$$

*So the force is 20 N everywhere.* □

### 17.1.2 Bar and Beam Elements

The spring system we discussed earlier is limited in many ways as a spring does not have any mass and its cross section is not explicitly included. A more complicated but realistic element is the bar element as shown in Figure 17.3, which is a uniform rod with a cross section area $A$, Young's elastic modulus $E$, and a length $L$. A bar element can only support tension and compression, it cannot support bending. For this reason, it is also called a truss element.

The displacements at nodes $i$ and $j$ are $u_i$ and $u_j$, respectively. The forces at the corresponding nodes are $f_i$ and $f_j$. Now we have to derive its stiffness matrix. Assuming the bar is linearly elastic, the stress $\sigma$ is thus related to strain $\epsilon$ via

$\sigma = E\epsilon$. Since $\epsilon = (u_j - u_i)/L$ and $\sigma = f/A$ where $F$ is the force in the bar element, we have

$$f = \frac{EA}{L}(\Delta u) = k(\Delta u), \tag{17.17}$$

where $\Delta u = u_j - u_i$ is the extension or elongation of the bar element. Now the equivalent spring stiffness constant is

$$k = \frac{EA}{L}. \tag{17.18}$$

Therefore, the stiffness matrix $\mathbf{K}$ for this bar becomes

$$\mathbf{K} = \begin{pmatrix} k & -k \\ -k & k \end{pmatrix} = \frac{EA}{L}\begin{pmatrix} 1 & -1 \\ -1 & 1 \end{pmatrix}. \tag{17.19}$$

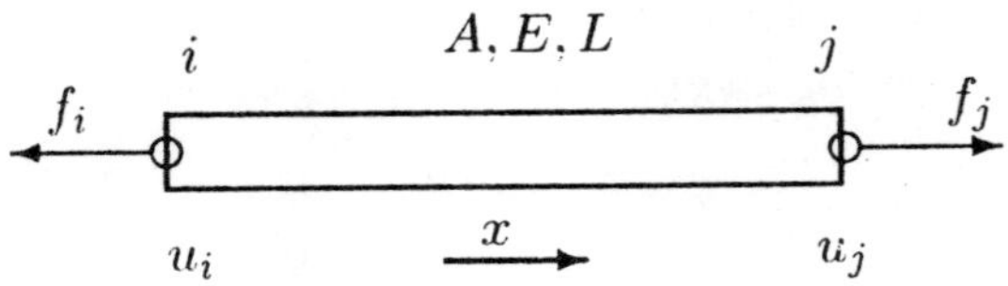


Figure 17.3: Bar element.

We have up to now only discussed 1-D systems where all displacements $u_i$ or $u_j$ are along the bar direction, and each node has only one displacement (one degree of freedom). We now extend to study 2-D systems. In 2-D, each node $i$ has two displacements $u_i$ (along the bar direction) and $v_i$ (perpendicular to the bar direction). Thus, each node has two degrees of freedom.

If we rotate the bar element by an angle $\theta$ as shown in Figure 17.4, we cannot use the standard addition to assemble the system. A transformation is needed between the global coordinates $(x, y)$ to the local coordinates $(x', y')$. From the geometrical consideration, the global displacements $u_i$ and $v_i$

at node $i$ are related to the local displacement $u'_i$ and (usually) $v'_i = 0$.

$$\begin{pmatrix} u'_i \\ v'_i \end{pmatrix} = \begin{pmatrix} \cos\theta & \sin\theta \\ -\sin\theta & \cos\theta \end{pmatrix} \begin{pmatrix} u_i \\ v_i \end{pmatrix}. \tag{17.20}$$

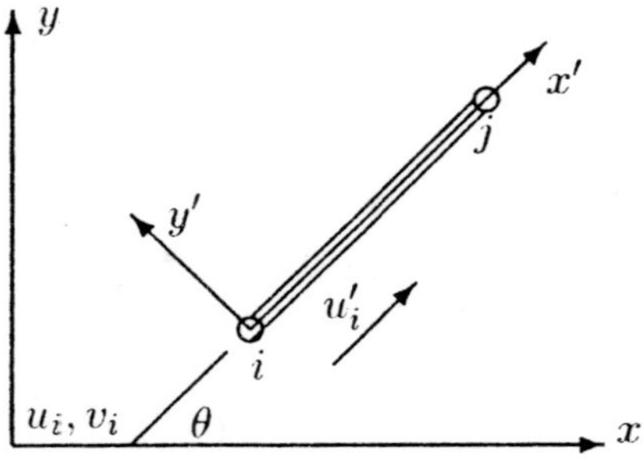


Figure 17.4: 2-D transformation of coordinates.

Using the similar transformation for $u_j$ and $v_j$, we get the transformation for the two-node bar element

$$\mathbf{u}' = \begin{pmatrix} u'_i \\ v'_i \\ u'_j \\ v'_j \end{pmatrix} = \begin{pmatrix} \cos\theta & \sin\theta & 0 & 0 \\ -\sin\theta & \cos\theta & 0 & 0 \\ 0 & 0 & \cos\theta & \sin\theta \\ 0 & 0 & -\sin\theta & \cos\theta \end{pmatrix} \begin{pmatrix} u_i \\ v_i \\ u_j \\ v_j \end{pmatrix}$$

which can be written as

$$\mathbf{u}' = \mathbf{R}\mathbf{u}, \tag{17.21}$$

where

$$\mathbf{R} == \begin{pmatrix} \cos\theta & \sin\theta & 0 & 0 \\ -\sin\theta & \cos\theta & 0 & 0 \\ 0 & 0 & \cos\theta & \sin\theta \\ 0 & 0 & -\sin\theta & \cos\theta \end{pmatrix}. \tag{17.22}$$

The same applies to transform the force,

$$\mathbf{f}' = \mathbf{R}\mathbf{f}, \tag{17.23}$$

and the stiffness matrix in local coordinates is

$$\mathbf{K}'\mathbf{u}' = \mathbf{f}'. \tag{17.24}$$

As the calculation is mainly based on the global coordinates, and the assembly should be done by transforming the local systems to the global coordinates, by combining the above two equations, we have

$$\mathbf{K}'\mathbf{R}\mathbf{u} = \mathbf{R}\mathbf{f}. \tag{17.25}$$

or

$$\mathbf{R}^{-1}\mathbf{K}'\mathbf{R}\mathbf{u} = \mathbf{K}\mathbf{u} = \mathbf{f}, \tag{17.26}$$

which is equivalent to a global stiffness matrix

$$\mathbf{K} = \mathbf{R}^{-1}\mathbf{K}'\mathbf{R}. \tag{17.27}$$

The stiffness matrix $\mathbf{K}$ is a $4 \times 4$ matrix in 2-D.

Bar elements can only elongate or shrink, they do not support bending or deflection. For bending, we need the beam elements which include a rotation around the end nodes $\theta_i$ and $\theta_j$. In this case, each node has three degrees of freedom $(u_i, v_i, \theta_i)$, and the stiffness matrix is therefore a $6 \times 6$ matrix in 2-D. For more complicated elements, it is necessary to use a formal approach in terms of shape functions and weak formulations.

## 17.2 Finite Element Formulation

### 17.2.1 Weak Formulation

Many problems are modelled in terms of partial differential equations, which can generally be written as

$$\mathcal{L}(u) = 0, \qquad \mathbf{x} \in \Omega, \tag{17.28}$$

where $\mathcal{L}$ is a differential operator, often linear. This problem is usually completed with the essential boundary condition $\mathcal{E}(u) = (u - \bar{u}) = 0$ for $\mathbf{x} \in \partial\Omega_E$ and natural boundary conditions $\mathcal{B}(u) = 0$ for $\mathbf{x} \in \partial\Omega_N$. Assuming that the solution can be approximated by $u_h$ over a finite element mesh with an averaged element size or mean distance $h$ between two adjacent nodes, the above equation can be approximated as

$$\mathcal{L}(u_h) \approx 0. \tag{17.29}$$

Multiplying both sides of the equation by a test function or a proper weighting function, integrating over the domain and using associated boundary conditions, we can write the general weak formulation of Zienkiewicz-type as

$$\int_\Omega \mathcal{L}(u_h) w_i d\Omega$$

$$+ \int_{\partial\Omega_N} \mathcal{B}(u_h)\bar{w}_i d\Gamma + \int_{\partial\Omega_E} \mathcal{E}(u_h)\tilde{w}_i d\Gamma_E \approx 0, \tag{17.30}$$

where $(i = 1, 2, ..., M)$. If we can approximate the solution $u_h$ by the expansion

$$u_h(u, t) = \sum_{i=1}^{M} u_i(t) N_i(x) = \sum_{j=1}^{M} u_j N_j, \tag{17.31}$$

it requires that $N_i = 0$ on $\partial\Omega_E$ so that $\tilde{w}_i = 0$ on $\partial\Omega_E$. Thus, only the natural boundary conditions are included since the essential boundary conditions are automatically satisfied. In addition, there is no much limitation on the choice of $w_i$ and $\bar{w}_i$. If we choose $\bar{w}_i = -w_i$ so as to simplify the formulation, we have

$$\int_\Omega \mathcal{L}(u_h) w_i d\Omega \approx \int_{\partial\Omega_N} \mathcal{B}(u_h) w_i d\Gamma. \tag{17.32}$$

### 17.2.2 Galerkin Method

There are many different ways to choose the test functions $w_i$ and shape functions $N_i$. One of the most popular methods is the Galerkin method where the test functions are the same as the shape functions, or $w_i = N_i$. In this special case, the formulation simply becomes

$$\int_\Omega \mathcal{L}(u_h) N_i d\Omega \approx \int_{\partial\Omega_N} \mathcal{B}(u_h) N_i d\Gamma. \tag{17.33}$$

The discretization of this equation will usually lead to an algebraic matrix equation.

On the other hand, if we use the Dirac delta function as the test functions $w_i = \delta(\mathbf{x} - \mathbf{x}_i)$, the method is called the collocation method which uses the interesting properties of the Dirac function

$$\int_\Omega f(\mathbf{x})\delta(\mathbf{x} - \mathbf{x}_i)d\Omega = f(\mathbf{x}_i), \tag{17.34}$$

together with $\delta(\mathbf{x} - \mathbf{x}_i) = 1$ at $\mathbf{x} = \mathbf{x}_i$ and $\delta(\mathbf{x} - \mathbf{x}_i) = 0$ at $\mathbf{x} \neq \mathbf{x}_i$.

### 17.2.3 Shape Functions

The main aim of the finite element method is to find an approximate solution $u_h(\mathbf{x}, t)$ for the exact solution $u$ on some nodal points,

$$u_h(\mathbf{x}, t) = \sum_{i=1}^{M} u_i(t)N_i(\mathbf{x}) \tag{17.35}$$

where $u_i$ are unknown coefficients or the value of $u$ at the discrete nodal point $i$. Functions $N_i$ $(i = 1, 2, ..., M)$ are linearly independent functions that vanish on the part of the essential boundary. At any node $i$, we have $N_i = 1$, and $N_i = 0$ at any other nodes, or

$$\sum_{i=1}^{M} N_i = 1, \qquad N_i(\mathbf{x}_j) = \delta_{ij}. \tag{17.36}$$

The functions $N_i(\mathbf{x})$ are referred to as basis functions, trial functions or more often shape functions in the literature of finite element methods. For the simplest 1-D element with two nodes $i$ and $j$, the linear shape functions can be written as

$$N_i = 1 - \xi = 1 - \frac{x}{L}, N_j = \xi = \frac{x}{L}, \tag{17.37}$$

where $L = |x_j - x_i|$, which is shown in Figure 17.5.

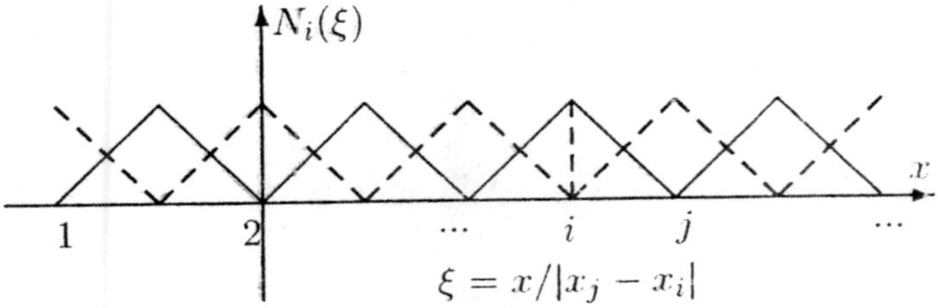


Figure 17.5: The 1-D linear shape functions.

Using the assumptions that $u_i(t)$ does not depend on space and $N_i(\mathbf{x})$ does not depend on time, the derivatives of $u$ can be approximated as

$$\frac{\partial u}{\partial \mathbf{x}} \approx \frac{\partial u_h}{\partial \mathbf{x}} = \sum_{i=1}^{M} u_i(t) N'(\mathbf{x}),$$

$$\dot{u} \approx \frac{\partial u_h}{\partial t} = \sum_{i=1}^{M} \dot{u}_i N(\mathbf{x}), \tag{17.38}$$

where we have used the notations: $' = d/d\mathbf{x}$ and $\dot{} = \frac{\partial}{\partial t}$. Higher order derivatives are then calculated in a similar way. The ultimate goal is to construct a method of computing $u_i$ such that the error $u_h - u$ is minimized. Generally speaking, the residual $R$ varies with space and time, so we have

$$R(u_1, ..., u_M, \mathbf{x}) = \mathcal{L}(u_h(\mathbf{x})). \tag{17.39}$$

There are several methods to minimize $R$. Depending on the scheme of minimization and the choice of shape functions, various methods can be formulated. These include the weighted residual method, the method of least squares, the Galerkin method and others.

## 17.3 Elasticity

### 17.3.1 Plane Stress and Plane Strain

The stress tensor $\boldsymbol{\sigma}$ and strain tensor $\boldsymbol{\epsilon}$ are not written as tensor forms but vector forms $\boldsymbol{\sigma} = (\sigma_{xx}, \sigma_{yy}, \sigma_{zz}, \sigma_{xy}, \sigma_{yz}, \sigma_{zx})^T$, and $\boldsymbol{\epsilon} = (\epsilon_{xx}, \epsilon_{yy}, \epsilon_{zz}, \gamma_{xy}, \gamma_{yz}, \gamma_{zx})^T$. The strain tensor is usually defined as

$$\varepsilon_{ij} = \frac{1}{2}\left(\frac{\partial u_i}{\partial x_j} + \frac{\partial u_j}{\partial x_i}\right), \tag{17.40}$$

where one applies the engineering shear strain $\epsilon_{xy} = 2\varepsilon_{xy}$. Hooke's elasticity can be expressed as

$$\boldsymbol{\sigma} = \mathbf{D}\boldsymbol{\epsilon}, \tag{17.41}$$

where $\mathbf{D}$ is a $6 \times 6$ symmetrical matrix as functions of Young's modulus $E$ and Poisson's ratio $\nu$.

Two special cases that are commonly found in many applications are the plane stress ($\sigma_{zz} = 0$, but $\epsilon_{zz} \neq 0$) and plane strain ($\epsilon_{zz} = 0$, but $\sigma_{zz} \neq 0$). The commonly used formulation is the displacement-based formulation or $u$-based formulation. In the 2-D case, the displacement $\mathbf{u} = (u, v)^T$ and the strain $\epsilon$ and stress $\sigma$ are defined as

$$\sigma = \begin{pmatrix} \sigma_x & \sigma_y & \tau_{xy} \end{pmatrix}^T, \quad \epsilon = \begin{pmatrix} \epsilon_x & \epsilon_y & \gamma_{xy} \end{pmatrix}^T. \tag{17.42}$$

Now the stress-strain relationship becomes

$$\sigma = \mathbf{D}(\epsilon - \epsilon_0), \tag{17.43}$$

where $\mathbf{D}$ is a $3 \times 3$ matrix. The strains are given by

$$\epsilon_x = \frac{\partial u}{\partial x}, \quad \epsilon_y = \frac{\partial v}{\partial y}, \quad \gamma_{xy} = \frac{\partial u}{\partial y} + \frac{\partial v}{\partial x}, \tag{17.44}$$

where $\epsilon_0$ is the initial strain due to temperature change or thermal loading. If there is no such change, then the initial strain can be taken to be zero in most applications.

The equilibrium of force in elasticity leads to

$$\nabla \cdot \sigma + \mathbf{b} = \mathbf{0}. \tag{17.45}$$

where $\mathbf{b} = [f_x \quad f_y]^T$ is the body force. In the case of plane stress, we have

$$\mathbf{D} = \frac{E}{1-\nu^2}\begin{pmatrix} 1 & \nu & 0 \\ \nu & 1 & 0 \\ 0 & 0 & (1-\nu)/2 \end{pmatrix}. \tag{17.46}$$

In the case of plane strain, we have

$$\mathbf{D} = \frac{E}{(1-2\nu)}\begin{pmatrix} \frac{1-\nu}{1+\nu} & \frac{\nu}{1+\nu} & 0 \\ \frac{\nu}{1+\nu} & \frac{1-\nu}{1+\nu} & 0 \\ 0 & 0 & \frac{(1-2\nu)}{2(1+\nu)} \end{pmatrix}. \tag{17.47}$$

Clearly, for the 1-D case plane stress when $v = 0$, $f_y = 0$ and $\sigma_y = \tau_{xy} = 0$, the equation of force balance simply becomes

$$\frac{E}{1-\nu^2}\frac{\partial^2 u}{\partial x^2} + f_x = 0, \tag{17.48}$$

where we have used the stress-strain relationship. This 1-D equation is essentially the same as the 1-D heat transfer equation $u'' + Q = 0$ that will be discussed in detail later, so the solution technique for the 1-D heat transfer shall equally apply. Therefore, we shall focus on the 2-D case in the rest of this section.

Displacements $(u, v)$ in a plane element can be interpolated from nodal displacements. For example, using a triangular element $(i, j, m)$ with three nodal points as shown in Figure 17.6 $(x_i, y_i)$, $(x_j, y_j)$, and $(x_m, y_m)$, we have

$$\mathbf{u} = \begin{pmatrix} u \\ v \end{pmatrix} = [N_i\mathbf{I}, N_j\mathbf{I}, N_m\mathbf{I}]\begin{pmatrix} u_i \\ v_i \\ u_j \\ v_j \\ u_m \\ v_m \end{pmatrix} = \mathbf{Nd}, \tag{17.49}$$

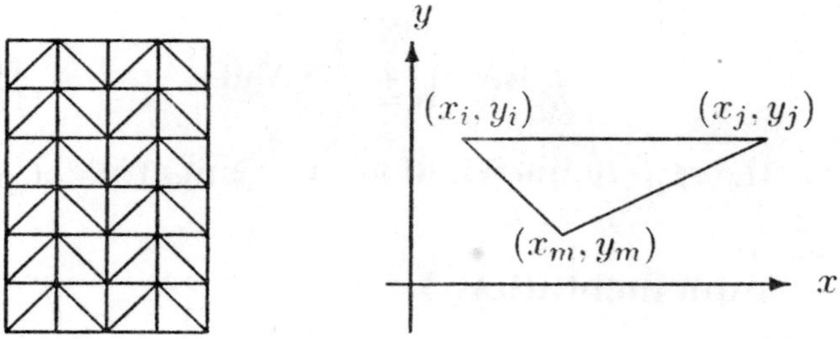

Figure 17.6: Schematic triangular mesh and the layout of a triangular element.

where $\mathbf{I}$ is a $2 \times 2$ unitary matrix, i.e.,

$$\mathbf{I} = \begin{pmatrix} 1 & 0 \\ 0 & 1 \end{pmatrix}. \tag{17.50}$$

and

$$\mathbf{N} = \begin{pmatrix} N_i & 0 & N_j & 0 & N_m & 0 \\ 0 & N_i & 0 & N_j & 0 & N_m \end{pmatrix}. \tag{17.51}$$

By defining a differential operator

$$\mathbf{L_d} = \begin{pmatrix} \frac{\partial}{\partial x} & 0 \\ 0 & \frac{\partial}{\partial y} \\ \frac{\partial}{\partial y} & \frac{\partial}{\partial x} \end{pmatrix}, \tag{17.52}$$

we can rewrite the above formulation as

$$\epsilon = \mathbf{L_d u} = \mathbf{L_d N d} = \mathbf{B d}, \tag{17.53}$$

where

$$\mathbf{B} = \mathbf{L_d N}. \tag{17.54}$$

Now the equation (17.45) becomes

$$\mathbf{K u} = \mathbf{f}, \tag{17.55}$$

where

$$\mathbf{K} = \int_\Omega \mathbf{B}^{\mathbf{T}} \mathbf{D} \mathbf{B} dV. \tag{17.56}$$

and

$$f_i = \int_{\Omega} \mathbf{b} N_i dV + \int_{\Gamma} \tau N_i d\Gamma, \tag{17.57}$$

where $\tau$ is the surface traction (force per unit area).

### 17.3.2 Implementation

In the case of 2-D elastic problems, the simplest elements are linear triangular elements, thus we have $\mathbf{B}_i = \mathbf{L}_d N_i$, or

$$\mathbf{B} = \mathbf{L}_d \mathbf{N} = \frac{1}{2\Delta} \begin{pmatrix} b_i & 0 & b_j & 0 & b_m & 0 \\ 0 & c_i & 0 & c_j & 0 & c_m \\ c_i & b_i & c_j & b_j & c_m & b_m \end{pmatrix}, \tag{17.58}$$

where

$$\Delta = \frac{1}{2} \begin{vmatrix} 1 & x_i & y_i \\ 1 & x_j & y_j \\ 1 & x_m & y_m \end{vmatrix} \tag{17.59}$$

is the area of the triangular element. The linear element implies that the strain is constant in the element. The stiffness matrix $\mathbf{K}_{ij}^{(e)} (i, j = 1, 2, ..., 6)$ of each triangular element can be expressed as

$$\mathbf{K}^{(e)} = \int_{\Omega_e} \mathbf{B}^T \mathbf{D} \mathbf{B} \ dxdy, \tag{17.60}$$

For a triangular element with three nodes $(i, j, m)$, each node has two degrees of freedom $(u_i, v_i)$, then the stiffness matrix $\mathbf{K}^{(e)}$ for each element is a $6 \times 6$ matrix. In general, if each element has $r$ nodes, and each node has $n$ degrees of freedom, then the local stiffness matrix is a $rn \times rn$ matrix. If the region has $M$ nodes in total, then $Mn$ equations are needed for this problem. For the present case ($n = 2$), we need $2M$ equations for plane stress and plane strain.

In order to calculate the contribution of each element to the overall (global) equation, each node should be identified in some way, and most often in terms of the index matrix. The nodal index matrix for three nodes $(i, j, m)$ can be written as

$$\mathrm{ID_{node}} = \begin{pmatrix} (i,i) & (i,j) & (i,m) \\ (j,i) & (j,j) & (j,m) \\ (m,i) & (m,j) & (m,m) \end{pmatrix}. \tag{17.61}$$

The nodal index matrix identifies the related nodes in the global node-numbering system. However, the main assembly of the stiffness matrix is about the corresponding equations and the application of the boundary conditions, thus we need to transfer the nodal index matrix to the equation index matrix in terms of the global numbering of equations with two degrees of freedom for each node (e.g., equations $2i-1$ and $2i$ for nodal $i$, equation $2j-1$ and $2j$ for node $j$, etc). Thus, for each entry in the stiffness matrix, say $(i, j)$, in the nodal index matrix, we now have four entries, i.e.,

$$(i,j) \rightarrow \begin{pmatrix} (2i-1, 2j-1) & (2i-1, 2j) \\ (2i, 2j-1) & (2i, 2j) \end{pmatrix}, \quad etc \tag{17.62}$$

The equation index matrix now has $6 \times 6$ entries, and each entry is a pair such as $(2i-1, 2j-1)$, ... , $(2m, 2m)$, etc. By writing it as two index matrices ($ID = JD^T$), we now have

$$\mathrm{ID_{equ}} = \mathrm{JD^T_{equ}},$$

$$\mathrm{JD_{equ}} = \begin{pmatrix} 2i-1 & 2i & 2j-1 & 2j & 2m-1 & 2m \\ 2i-1 & 2i & 2j-1 & 2j & 2m-1 & 2m \\ 2i-1 & 2i & 2j-1 & 2j & 2m-1 & 2m \\ 2i-1 & 2i & 2j-1 & 2j & 2m-1 & 2m \\ 2i-1 & 2i & 2j-1 & 2j & 2m-1 & 2m \\ 2i-1 & 2i & 2j-1 & 2j & 2m-1 & 2m \end{pmatrix} \tag{17.63}$$

So that the contribution of $K^{(e)}_{ij}$ to the global matrix $K_{ij}$ is simply

$$K_{[ID_{equ}(I,J), JD_{equ}(I,J)]} = K_{[ID_{equ}(I,J), JD_{equ}(I,J)]} + K^{(e)}_{(I,J)},$$

$$I, J = 1, 2, ..., 6. \tag{17.64}$$

Similarly, the contribution of the body force and external force can be computed

$$f(l) = f(l) + f^{(e)}(l)$$

where $l = 2i-1, 2i, 2j-1, 2j, 2m-1, 2m$ etc.

## 17.4 Heat Conduction

Heat transfer problems are very common in engineering and the geometry in most applications is irregular. Thus, finite element methods are especially useful in this case.

### 17.4.1 Basic Formulation

The steady-state heat transfer is governed by the heat conduction equation or Poisson's equation

$$\nabla \cdot (k \nabla u) + Q = 0, \tag{17.65}$$

with the essential boundary condition

$$u = \bar{u}, \qquad \mathbf{x} \in \partial\Omega_E, \tag{17.66}$$

and the natural boundary condition

$$k\frac{\partial u}{\partial n} - q = 0, \qquad \mathbf{x} \in \partial\Omega_N. \tag{17.67}$$

Using the formulation similar to the formulation (17.33) in terms of $u \approx u_h$, we have

$$\int_{\Omega} [\nabla \cdot (k\nabla u) + Q] N_i d\Omega - \int_{\partial\Omega_N} [k\frac{\partial u}{\partial n} - q] N_i d\Gamma = 0. \tag{17.68}$$

Integrating by parts and using Green's theorem, we have

$$-\int_{\Omega} (\nabla u_h \cdot k \cdot \nabla N_i) d\Omega + \int_{\partial\Omega} k\frac{\partial u_h}{\partial n} N_i d\Gamma$$

$$+\int_\Omega QN_i d\Omega - \int_{\partial\Omega_N} [k\frac{\partial u_h}{\partial n} - q]N_i d\Gamma = 0. \quad (17.69)$$

Since $N_i = 0$ on $\partial\Omega_E$, thus we have

$$\int_{\partial\Omega} [\,]N_i d\Gamma = \int_{\partial\Omega_N} [\,]N_i d\Gamma. \quad (17.70)$$

Therefore, the above weak formulation becomes

$$\int_\Omega (\nabla u_h \cdot k \cdot \nabla N_i) d\Omega - \int_\Omega QN_i d\Omega - \int_{\partial\Omega_N} qN_i d\Gamma = 0. \quad (17.71)$$

Substituting $u_h = \sum_{j=1}^{M} u_j N_j(\mathbf{x})$ into the equation, we have

$$\sum_{j=1}^{M} [\int_\Omega (k\nabla N_i \cdot \nabla N_j) d\Omega] u_j - \int_\Omega QN_i d\Omega$$

$$-\int_{\partial\Omega_N} qN_i d\Gamma = 0. \quad (17.72)$$

This can be written in the compact matrix form

$$\sum_{j=1}^{M} K_{ij} U_j = f_i, \qquad \mathbf{KU} = \mathbf{f}, \quad (17.73)$$

where $\mathbf{K} = [K_{ij}], (i, j = 1, 2, ..., M)$, $\mathbf{U}^T = (u_1, u_2, ..., u_M)$, and $\mathbf{f}^T = (f_1, f_2, ..., f_M)$. That is,

$$K_{ij} = \int_\Omega k\nabla N_i \nabla N_j d\Omega, \quad (17.74)$$

$$f_i = \int_\Omega QN_i d\Omega + \int_{\partial\Omega_N} qN_i d\Gamma. \quad (17.75)$$

---

□ ***Example 17.2(a):*** *As a simple example, we consider the 1-D steady-state heat conduction problem,*

$$u''(x) + Q(x) = 0,$$

*with boundary conditions*

$$u(0) = \beta, \quad u'(1) = q.$$

*For a special case* $Q(x) = r\exp(-x)$, *we have the analytical solution*

$$u(x) = (\beta - r) + (re^{-1} + q)x + re^{-x}. \qquad (17.76)$$

*Then equation (17.75) becomes*

$$\sum_{j=1}^{M}\Big(\int_0^1 N_i' N_j' dx\Big)u_j = \int_0^1 QN_i dx + qN_i(1).$$

*For the purpose of demonstrating the implementation procedure, let us solve this problem by dividing the interval into 4 elements and 5 nodes. This will be discussed later in more details.* □

## 17.4.2 Element-by-Element Assembly

The assembly of the linear matrix system is the popular element-by-element method. The stiffness matrix $\mathbf{K}$ in equations (17.73) and (17.75) is the summation of the integral over the whole solution domain, and the domain is now divided into $m$ elements with each element on a subdomain $\Omega_e$ ($e = 1, 2, ..., m$). Each element contributes to the whole stiffness matrix, and in fact, its contribution is a pure number. Thus, assembly of the stiffness matrix can be done in an element-by-element manner. Furthermore, $K_{i,j} \neq 0$ if and only if (or *iff*) nodes $i$ and $j$ belong to the same elements. In the 1-D case, $K_{i,j} \neq 0$ only for $j = i-1, i, i+1$. In finite element analysis, the shape functions $N_j$ are typically localized functions, thus the matrix $\mathbf{K}$ is usually sparse in most cases.

The element-by-element formulation can be written as

$$K_{i,j} = \sum_{e=1}^{m} K_{i,j}^{(e)}, \qquad K_{i,j}^{(e)} = \int_{\Omega_e} k\nabla N_i \nabla N_j d\Omega_e, \qquad (17.77)$$

and

$$f_i = \sum_{e=1}^{m} f_i^{(e)}, \qquad f_i^{(e)} = \int_{\Omega_e} QN_i d\Omega_e + \int_{\partial\Omega_{Ne}} qN_i d\Gamma_e. \quad (17.78)$$

In addition, since the contribution of each element is a simple number, the integration of each element can be done using the local coordinates and local node numbers or any coordinate system for the convenience of integration over an element. Then, the nonzero contribution of each element to the global system matrix **K** is simply assembled by direct addition to the corresponding global entry (of the stiffness matrix) of the corresponding nodes or related equations. In reality, this can be easily done using an index matrix to trace the element contribution to the global system matrix.

---

□ ***Example 17.2(b):*** *The assembly of the global system matrix for the example with 4 elements and five nodes is shown below. For each element with $i$ and $j$ nodes, we have*

$$N_i = 1 - \xi, \quad N_j = \xi, \quad \xi = \frac{x}{L}, \quad L = h_e,$$

$$K_{ij}^{(e)} = [\int_0^L k N_i' N_j' dx] = \frac{k}{h_e}\begin{pmatrix} 1 & -1 \\ -1 & 1 \end{pmatrix},$$

$$f_i^{(e)} = \frac{Qh_e}{2}\begin{pmatrix} 1 \\ 1 \end{pmatrix},$$

*so that, for example in elements 1 and 2, these can extend to all nodes (with $h_i = x_{i+1} - x_i, i = 1, 2, 3, 4$),*

$$K^{(1)} = \begin{pmatrix} k/h_1 & -k/h_1 & 0 & 0 & 0 \\ -k/h_1 & k/h_1 & 0 & 0 & 0 \\ 0 & 0 & 0 & 0 & 0 \\ 0 & 0 & 0 & 0 & 0 \\ 0 & 0 & 0 & 0 & 0 \end{pmatrix}, \quad f^{(1)} = \frac{Q}{2}\begin{pmatrix} h_1 \\ h_1 \\ 0 \\ 0 \\ 0 \end{pmatrix},$$

$$K^{(2)} = \begin{pmatrix} 0 & 0 & 0 & 0 & 0 \\ 0 & k/h_2 & -k/h_2 & 0 & 0 \\ 0 & -k/h_2 & k/h_2 & 0 & 0 \\ 0 & 0 & 0 & 0 & 0 \\ 0 & 0 & 0 & 0 & 0 \end{pmatrix}, \quad f^{(2)} = \frac{Q}{2}\begin{pmatrix} 0 \\ h_2 \\ h_2 \\ 0 \\ 0 \end{pmatrix},$$

*and so on. Now the global system matrix becomes*

$$K =$$

$$\begin{pmatrix} k/h_1 & -k/h_1 & 0 & 0 & 0 \\ -k/h_1 & k/h_1 + k/h_2 & -k/h_2 & 0 & 0 \\ 0 & -k/h_2 & k/h_2 + k/h_3 & -k/h_3 & 0 \\ 0 & 0 & -k/h_3 & k/h_3 + k/h_4 & -k/h_4 \\ 0 & 0 & 0 & -k/h_4 & k/h_4 \end{pmatrix},$$

$$\mathbf{U} = \begin{pmatrix} u_1 \\ u_2 \\ u_3 \\ u_4 \\ u_5 \end{pmatrix}, \quad \boldsymbol{f} = \begin{pmatrix} Qh_1/2 \\ Q(h_1 + h_2)/2 \\ Q(h_2 + h_3)/2 \\ Q(h_3 + h_4)/2 \\ Qh_4/2 + q \end{pmatrix},$$

*where the last row of* $\boldsymbol{f}$ *has already included the natural boundary condition at* $u'(1) = q$. □

### 17.4.3 Application of Boundary Conditions

Boundary conditions can be essential, natural or mixed. The essential boundary conditions are automatically satisfied in the finite element formulation by the approximate solution. These include the displacement, rotation, and known value of the solution. Sometimes, they are also called the geometric boundary conditions. In our example, it is $u(0) = \beta$. Natural boundary conditions often involve the first derivatives such as strains, heat flux, force, and moment. Thus, they are also referred to as force boundary conditions. In our example, it is $u'(1) = q$.

The natural boundary conditions are included in the integration in the finite element equations such as (17.75). Thus no further imposition is necessary. On the other hand, although the essential boundary conditions are automatically satisfied in the finite element formulations, they still need to be implemented in the assembled finite element equations to ensure unique solutions. The imposition of the essential boundary conditions can be done in main several ways: a) direct application; b) Lagrangian multiplier and c) penalty method. To show how these methods work, we use the 1-D poisson equation on the distinct points $x_i (i = 1, 2, ..., M) \in [0, 1]$ to aid our discussion.

### Direct Application

In this method, we simply use the expansion $u_h = \sum_{i=1}^{M} u_i N_i$, and apply directly the essential boundary conditions at point $i$ to replace the corresponding $i$th equation with $u_i = \hat{u}_i$ so that $i$th row of the stiffness matrix $\mathbf{K}$ in equation (17.73) becomes $(0, 0, ..., 1, ..., 0)$ and the corresponding $f_i = f(i) = \hat{u}_i$. All other points will be done in the similar manner. For example, the boundary conditions $u(0) = \alpha$ and $u(M) = \beta$ in the 1-D case mean that the first and last equations are replaced by $u_1 = \alpha$ and $u_M = \beta$, respectively. Thus, $K_{11} = 1, f_1 = \alpha$ (all other coefficients are set to be zeros: $K_{12} = ... = K_{1M} = 0$, and $K_{MM} = 1, f_M = \beta$ with $K_{M1} = ... = K_{M,M-1} = 0$). Then, the equations can be solved for $(u_1, u_2, ..., u_M)^T$. This method is widely used due to its simplicity and the advantage of time-stepping because it allows bigger time steps.

### Lagrangian Multiplier

This method is often used in the structure and solid mechanics to enforce the constraints $(u_i = \bar{u}_i)$. The variation is added by the extra term $\lambda(u_i - \bar{u}_i)$ where $\lambda$ is the Lagrange multiplier. Now we have

$$\Pi = \frac{1}{2}\mathbf{u^T K u} - \mathbf{u^T f} + \lambda(u_i - \bar{u}_i), \tag{17.79}$$

whose variation $\delta\Pi = 0$ leads to

$$\delta\mathbf{u^T K u} - \delta\mathbf{u^T f} + \lambda\delta u_i + \delta\lambda(u_i - \bar{u}_i) = 0. \tag{17.80}$$

Because $\delta\mathbf{u}$ and $\delta\lambda$ are arbitrary, we have

$$\begin{pmatrix} \mathbf{K} & \boldsymbol{e}_i \\ \boldsymbol{e}_i^T & 0 \end{pmatrix} \begin{pmatrix} \mathbf{u} \\ \lambda \end{pmatrix} = \begin{pmatrix} \mathbf{f} \\ \bar{u}_i \end{pmatrix}.$$

where $\boldsymbol{e}_i = (0, 0, ..., 1, 0, ..., 0)^T$ (its $i$th entry is equal to one). This method can be extended to $m$ Lagrangian multipliers.

## Penalty Method

One of the most widely used methods of enforcing the essential boundary conditions is the so-called penalty method in terms of a very large coefficient $\gamma$, $u_i = \hat{u}$ at $\mathbf{x}_i \in \partial\Omega_E$, so that $\gamma u_i = \gamma \hat{u}$ can be directly added onto $\mathbf{Ku} = \mathbf{f}$. In the 1-D example, it simply leads to $K_{11} = K_{11} + \gamma$, $K_{MM} = K_{MM} + \gamma$, and $f_1 = f_1 + \gamma\alpha$, $f_M = f_M + \gamma\beta$. The common rule for choosing $\gamma$ is that $\gamma \gg \max|K_{ii}|$. Usually, $\gamma \approx 1000 \max|K_{ii}|$ should be adequate. The penalty method is widely used in steady-state problems. However, it may affect the efficiency of time-stepping since it increases the maximum eigenvalue of the stiffness matrix, and thus very small time steps are required for convergence. The advantage of the penalty method is that the handling of the essential boundary conditions becomes simpler from the implementation point of view. The disadvantage is that the conditions are only satisfied approximately.

---

□ ***Example 17.2(c):*** *Following the same example of the 1-D steady state heat conduction discussed earlier, we now use the direct application method for the essential boundary conditions. We can replace the first equation* $\sum_{j=1}^{5} K_{1j}u_j = f_1$ *with* $u_1 = \beta$, *so that the first row becomes* $K_{1j} = (1\ 0\ 0\ 0\ 0)$ *and* $f_1 = \beta$. *Thus, we have*

$$\mathbf{K} =$$

$$\begin{pmatrix} 1 & 0 & 0 & 0 & 0 \\ -k/h_1 & k/h_1 + k/h_2 & -k/h_2 & 0 & 0 \\ 0 & -k/h_2 & k/h_2 + k/h_3 & -k/h_3 & 0 \\ 0 & 0 & -k/h_3 & k/h_3 + k/h_4 & -k/h_4 \\ 0 & 0 & 0 & -k/h_4 & k/h_4 \end{pmatrix},$$

$$\mathbf{U} = \begin{pmatrix} u_1 \\ u_2 \\ u_3 \\ u_4 \\ u_5 \end{pmatrix}, \quad \boldsymbol{f} = \begin{pmatrix} \beta \\ Q(h_1 + h_2)/2 \\ Q(h_2 + h_3)/2 \\ Q(h_3 + h_4)/2 \\ Qh_4/2 + q \end{pmatrix},$$

*For the case of* $k = 1, Q = -1, h_1 = ... = h_4 = 0.25$, $\beta = 1$ *and*

$q = -0.25$, we have

$$\mathbf{K} = \begin{pmatrix} 1 & 0 & 0 & 0 & 0 \\ -4 & 8 & -4 & 0 & 0 \\ 0 & -4 & 8 & -4 & 0 \\ 0 & 0 & -4 & 8 & -4 \\ 0 & 0 & 0 & -4 & 4 \end{pmatrix}, \quad \boldsymbol{f} = \begin{pmatrix} 1 \\ -0.25 \\ -0.25 \\ -0.25 \\ -0.375 \end{pmatrix}$$

*Hence, the solution is*

$$\mathbf{U} = \mathbf{K}^{-1}\boldsymbol{f} = \begin{pmatrix} 1.00 \\ 0.72 \\ 0.50 \\ 0.34 \\ 0.25 \end{pmatrix}.$$

□

## 17.5 Time-Dependent Problems

The problems we have discussed so far in our finite element analysis are not time-dependent, and the solutions obtained are the steady-state solutions. However, most realistic problems involve time, and thus we will now discuss the time-dependent problems.

### 17.5.1 The Time Dimension

As the weak formulation uses the Green theorem that involves the spatial derivatives, the time derivatives can be considered as the source term. Thus, one simple and yet instructive way to extend the finite element formulation to include the time dimension is to replace $Q$ in equation (17.65) with $Q - \alpha u_t - \beta u_{tt} = Q - \alpha \dot{u} - \beta \ddot{u}$ so that we have

$$\nabla \cdot (k \nabla u) + (Q - \alpha u_t - \beta u_{tt}) = 0. \tag{17.81}$$

The boundary conditions and initial conditions are $u(\mathbf{x}, 0) = \phi(\mathbf{x})$, $u = \bar{u}, \mathbf{x} \in \partial\Omega_E$, and $k\frac{\partial u}{\partial n} - q = 0, \mathbf{x} \in \partial\Omega_N$. Using integration by parts and the expansion $u_h = \sum_{j=1}^{M} u_j N_j$, we have

$$\sum_{j=1}^{M}[\int_{\Omega}(k\nabla N_i \nabla N_j)d\Omega] + \sum_{j=1}^{M}\int_{\Omega}[(N_i \alpha N_j)\dot{u}_j + (N_i \beta N_j)\ddot{u}_j]d\Omega$$

$$-\int_{\Omega} N_i Q d\Omega - \int_{\partial\Omega_N} N_i q d\Gamma = 0, \tag{17.82}$$

which can be written in a compact form as

$$\mathbf{M}\ddot{\mathbf{u}} + \mathbf{C}\dot{\mathbf{u}} + \mathbf{K}\mathbf{u} = \mathbf{f}, \tag{17.83}$$

where

$$K_{ij} = \int_{\Omega}[(k\nabla N_i \nabla N_j)]d\Omega, \tag{17.84}$$

$$f_i = \int_{\Omega} N_i Q d\Omega + \int_{\partial\Omega_N} N_i q d\Gamma, \tag{17.85}$$

and

$$C_{ij} = \int_{\Omega} N_i \alpha N_j d\Omega, \qquad M_{ij} = \int_{\Omega} N_i \beta N_j d\Omega. \tag{17.86}$$

The matrices $\mathbf{K}, \mathbf{M}, \mathbf{C}$ are symmetric, that is to say, $K_{ij} = K_{ji}, M_{ij} = M_{ji}, C_{ij} = C_{ji}$ due to the interchangeability of the orders in the product of the integrand $k$, $N_i$ and $N_j$ (i.e., $\nabla N_i \cdot k \cdot \nabla N_j = k\nabla N_i \nabla N_j$, $N_i \alpha N_j = N_j \alpha N_i = \alpha N_i N_j$ etc). The matrix $\mathbf{C} = [C_{ij}]$ is the damping matrix similar to the damping coefficient of damped oscillations. $\mathbf{M} = [M_{ij}]$ is the general mass matrix due to a similar role acting as an equivalent mass in dynamics. In addition, before the boundary conditions are imposed, the matrix is usually singular, which may imply many solutions. Only after the proper boundary conditions have been enforced, the stiffness matrix will be nonsingular, thus unique solutions may be obtained. On the other hand, $\mathbf{M}$ and $\mathbf{C}$ will be always non-singular if they are not zero. For example, for the 1-D elements (with nodes $i$ and $j$),

$$K_{ij}^{(e)} = \frac{k}{h_e}\begin{pmatrix} 1 & -1 \\ -1 & 1 \end{pmatrix}, \qquad \det[K^{(e)}] = 0, \tag{17.87}$$

but

$$M_{ij}^{(e)} = \frac{\beta h_e}{6}\begin{pmatrix} 2 & 1 \\ 1 & 2 \end{pmatrix}, \qquad \det[M^{(e)}] \neq 0,$$

$$C_{ij}^{(e)} = \frac{\alpha h_e}{6}\begin{pmatrix} 2 & 1 \\ 1 & 2 \end{pmatrix}, \qquad \det[C^{(e)}] \neq 0, \tag{17.88}$$

Clearly, if $\mathbf{M} = 0$, it reduces to the linear heat conduction. If $\mathbf{C} = 0$, it becomes the wave equation with the source term.

### 17.5.2 Time-Stepping

From the general governing equation

$$\mathbf{M}\ddot{\mathbf{u}} + \mathbf{C}\dot{\mathbf{u}} + \mathbf{K}\mathbf{u} = \mathbf{f}, \tag{17.89}$$

we see that it is an ordinary differential equation in terms of time and matrices. Thus, in principle, all the time-stepping methods developed in the standard finite difference method can be used for this purpose. For a simple center difference scheme, we have

$$\dot{\mathbf{u}} = \frac{\mathbf{u}^{n+1} - \mathbf{u}^n}{\Delta t}, \qquad \ddot{\mathbf{u}} = \frac{(\mathbf{u}^{n+1} - 2\mathbf{u}^n + \mathbf{u}^{n-1})}{(\Delta t)^2}. \tag{17.90}$$

so that equation (17.89) becomes

$$\mathbf{M}\frac{(\mathbf{u}^{n+1} - 2\mathbf{u}^n + \mathbf{u}^{n-1})}{(\Delta t)^2}$$

$$+\mathbf{C}\frac{(\mathbf{u}^{n+1} - \mathbf{u}^{n-1})}{2\Delta t} + \mathbf{K}\mathbf{u}^n = \mathbf{f}. \tag{17.91}$$

Now the aim is to express $\mathbf{u}^{n+1}$ in terms of $\mathbf{u}^n$ and $\mathbf{u}^{n-1}$.

### 17.5.3 1-D Transient Heat Transfer

In the case of heat conduction ($\mathbf{M} = 0$), we have

$$\mathbf{C}\dot{\mathbf{u}} + \mathbf{K}\mathbf{u} = \mathbf{f}. \tag{17.92}$$

or

$$\dot{\mathbf{u}} = \mathbf{C}^{-1}(\mathbf{f} - \mathbf{K}\mathbf{u}). \tag{17.93}$$

Using the explicit time-stepping method, we can write it as

$$\frac{\mathbf{u}^{n+1} - \mathbf{u}^n}{\Delta t} = \mathbf{C}^{-1}(\mathbf{f} - \mathbf{K}\mathbf{u}^n), \tag{17.94}$$

so that we have

$$\mathbf{u}^{n+1} = \mathbf{u}^n + \Delta t \mathbf{C}^{-1}(\mathbf{f} - \mathbf{K}\mathbf{u}^n). \tag{17.95}$$

---

□ ***Example 17.3:*** *For a transient heat conduction problem, we have*

$$\alpha u_t = k u_{xx} + Q,$$

*and*

$$u(x,0) = 0, \quad u(0,t) = 1, \quad u'(1) = q.$$

*The formulation with 5 nodes and 4 elements leads to*

$$\mathbf{C} = \frac{\alpha}{6}\begin{pmatrix} 2h_1 & h_1 & 0 & 0 & 0 \\ h_1 & 2(h_1+h_2) & h_2 & 0 & 0 \\ 0 & h_2 & 2(h_2+h_3) & h_3 & 0 \\ 0 & 0 & h_3 & 2(h_3+h_4) & h_4 \\ 0 & 0 & 0 & h_4 & 2h_4 \end{pmatrix}$$

*For the case of $\alpha = 6, k = 1, Q = -1, h_1 = ... = h_4 = 0.25$, we have*

$$\mathbf{C} = \begin{pmatrix} 0.5 & 0.25 & 0 & 0 & 0 \\ 0.25 & 1 & 0.25 & 0 & 0 \\ 0 & 0.25 & 1 & 0.25 & 0 \\ 0 & 0 & 0.25 & 1 & 0.25 \\ 0 & 0 & 0 & 0.25 & 0.5 \end{pmatrix}.$$

□

---

### 17.5.4 Wave Equation

For the wave equation ($\mathbf{C} = 0$), we have

$$\mathbf{M}\ddot{\mathbf{u}} + \mathbf{K}\mathbf{u} = \mathbf{f}. \tag{17.96}$$

Using

$$\ddot{\mathbf{u}} = \frac{\mathbf{u}^{n+1} - 2\mathbf{u}^n + \mathbf{u}^{n-1}}{(\Delta t)^2}, \tag{17.97}$$

we have

$$\mathbf{u}^{n+1} = \mathbf{M}^{-1}\mathbf{f}(\Delta t)^2 + [2\mathbf{I} - (\Delta t)^2\mathbf{M}^{-1}\mathbf{K}]\mathbf{u}^n - \mathbf{u}^{n-1}, \tag{17.98}$$

where $\mathbf{I}$ is an identity or unit matrix. For example, the 1-D wave equation

$$\frac{\partial^2 u}{\partial t^2} = c\frac{\partial^2 u}{\partial x^2}, \tag{17.99}$$

with the boundary conditions

$$u(0) = u(1) = 0, \qquad u(x, 0) = e^{-(x-1/2)^2}, \tag{17.100}$$

can be written as

$$M_{ij} = \int_0^1 N_i N_j dx, \ \ K_{ij} = \int_0^1 cN_i' N_j' dx, \ \ \mathbf{f} = 0, \tag{17.101}$$

and $\mathbf{u}^0$ is derived from the $u(x, 0) = \exp[-(x - 1/2)^2]$.

The finite element methods in this book are mainly for linear partial differential equations. Although these methods can in principle be extended to nonlinear problems, however, some degrees of approximations and linearization are needed. In addition, an iterative procedure is required to solve the resultant nonlinear matrix equations. The interested readers can refer to many excellent books on these topics.

# Chapter 18

# Reaction Diffusion System

The partial differential equations we solved in the previous chapters using the three major numerical methods are linear equations. We know that the generalized forms of parabolic equations are nonlinear reaction-diffusion equations. Mathematically speaking, nonlinear equations are far more difficult to analyse if it is not impossible. From the numerical point of view, some extra linearization and approximations should be used for the nonlinear terms. However, the finite difference scheme should still be useful for most nonlinear equations though they should be implemented more carefully. Before we proceed to study the nonlinear system, let us review the fundamental characteristics of linear parabolic equations by solving the linear heat conduction equations.

## 18.1 Heat Conduction Equation

### 18.1.1 Fundamental Solutions

From the similarity solution in section 10.3, we know that both diffusion equation and heat conduction equation may mathematically have a similarity variable defined by

$$\zeta = \frac{x}{\sqrt{4kt}}, \tag{18.1}$$

where $k$ is either the thermal diffusivity or diffusion coefficient. In engineering, the coefficient $k$ has the unit $[\mathrm{m}]^2/[\mathrm{s}]$, thus the unit of $kt$ is $[\mathrm{m}]^2$, which means the variable $\zeta$ is dimensionless. Any two combinations of $x$ and $kt$ giving the same $\zeta$ will have similar solutions.

Using the similarity variable $\zeta$, the heat conduction equation can be transformed into an ordinary differential equation

$$f''(\zeta) = -2\zeta f', \qquad \text{or} \qquad (\ln f')' = -2\zeta. \tag{18.2}$$

Integrating it once, we have

$$\ln f' = Ce^{-\zeta^2}, \tag{18.3}$$

where $C$ is an integration constant. Integrating it again, we have

$$u = C \int_x^{x_0} e^{-\zeta^2} d\zeta + D, \tag{18.4}$$

which is the general solution of the heat conduction equation. If the domain is semi-infinite or infinite so that $x_0 \to 0$, then we get

$$= A\,\mathrm{erf}(\frac{x}{\sqrt{4kt}}) + B. \tag{18.5}$$

---

□ ***Example 18.1:*** *For the heat conduction in a semi-infinite domain, we have*

$$\frac{\partial u}{\partial t} = k\frac{\partial^2 u}{\partial x^2},$$

*with the boundary condition*

$$u = u_0, \qquad (x \leq 0),$$

*and an initial condition*

$$u = 0, \qquad (x > 0) \quad \text{at} \quad t = 0.$$

*The general solution is*

$$u = A\text{erf}(\frac{x}{\sqrt{4kt}}) + B.$$

*For* $x \to \infty$, $\text{erf}(x/\sqrt{4kt}) \to 1$, *we have*

$$A + B = 0.$$

*At* $x = 0$, $\text{erf}(0) = 0$, *we get* $B = u_0$. *The solution is*

$$u = u_0[1 - \text{erf}(\frac{x}{\sqrt{4kt}})] = u_0\text{erfc}\frac{x}{\sqrt{4kt}}.$$

*However, if* $u$ *is constant* $(u = u_0)$ *in the initial region* $x \in [-h, h]$, *then we have*

$$u = \frac{u_0}{2}[\text{erf}\frac{h-x}{\sqrt{4kt}} + \text{erf}\frac{h+x}{\sqrt{4kt}}].$$

□

The solutions of heat conduction do not always involve the error function because error functions only occur when the integration involves semi-infinite or infinite domains. If the domain has a finite length, then the solutions often consist of power series or even special functions. For example in heat conduction through a plane sheet with zero initial temperature, its two surfaces are held at constant temperatures with the boundary conditions $u = u_0$ at $x = 0$ for $(t \geq 0)$, and $u = 0$ at $x = L$ for $(t \geq 0)$. The general solution can be written as

$$u = u_0(1 - \frac{x}{L}) + \frac{2}{\pi}\sum_{n=1}^{\infty}\frac{u_0}{n}\sin\frac{n\pi x}{L}e^{-kn^2\pi^2 t/L^2}, \qquad (18.6)$$

which is a slowly convergent series.

## 18.2 Nonlinear Equations

### 18.2.1 Travelling Wave

The nonlinear reaction-diffusion equation

$$\frac{\partial u}{\partial t} = D\frac{\partial^2 u}{\partial x^2} + f(u). \qquad (18.7)$$

can have the travelling wave solution under appropriate conditions of $f(0) = f(1) = 0$, $f(u) > 0$ for $u \in [0, 1]$, and $f'(0) > 0$. For example, $f(u) = \gamma u(1 - u)$ satisfies these conditions, and the equation in this special case is called the Kolmogorov-Petrovskii-Piskunov (KPP) equation. By assuming that the travelling wave solution has the form $u(\zeta)$ and $\zeta = x - vt$, and substituting into the above equation, we have

$$Du''(\zeta) + vu'(\zeta) + f(u(\zeta)) = 0. \tag{18.8}$$

This is a second-order ordinary differential equation that can be solved with the appropriate boundary conditions

$$u(-\infty) \to 1, \qquad u(\infty) \to 0. \tag{18.9}$$

The KPP theory suggests that the limit of the speed of the travelling wave satisfies

$$v \geq 2\sqrt{Df'(0)}. \tag{18.10}$$

### 18.2.2 Pattern Formation

One of the most studied nonlinear reaction-diffusion equations in the 2-D case is the Kolmogorov-Petrovskii-Piskunov (KPP) equation

$$\frac{\partial u}{\partial t} = D(\frac{\partial^2 u}{\partial x^2} + \frac{\partial^2 u}{\partial y^2}) + \gamma q(u), \tag{18.11}$$

and

$$q(u) = u(1 - u). \tag{18.12}$$

The KPP equation can describe a huge number of physical, chemical and biological phenomena. The most interesting feature of this nonlinear equation is its ability of generating beautiful patterns. We can solve it using the finite difference scheme by applying the periodic boundary conditions and using a random initial condition $u = \text{random}(n, n)$ where $n$ is the size of the grid.

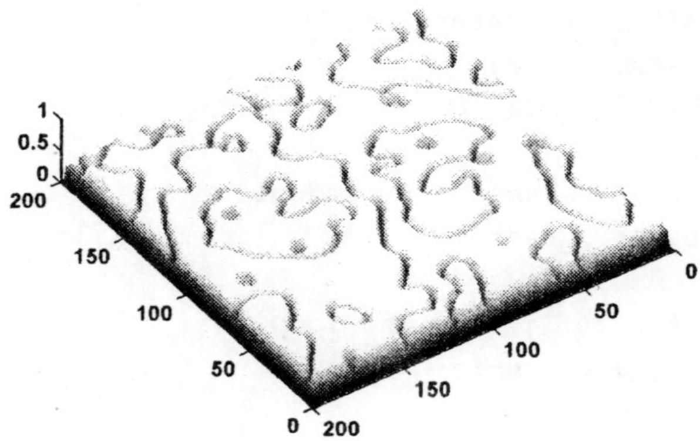


Figure 18.1: 2-D pattern formation for $D = 0.2$ and $\gamma = 0.5$.

Figure 18.1 shows the pattern formation of the above equation on a $100 \times 100$ grid for $D = 0.2$ and $\gamma = 0.5$. We can see that rings and thin curves are formed, arising from the random initial condition. The landscape surface shows the variations in the location and values of the field $u(x, y)$ can be easily demonstrated.

The following simple 15-line Matlab program can be used to solve this nonlinear system.

```
% Pattern formation:  a 15 line matlab program
% PDE form: u_t=D*(u_{xx}+u_{yy})+gamma*q(u)
% where q(u)='u.*(1-u)';
% The solution of this PDE is obtained by the
% finite difference method, assuming dx=dy=dt=1
% Written by X S Yang (Cambridge University)
% Usage: pattern(100)   or   simply  >pattern

function pattern(time)                      % line 1
% Input number of time steps
if nargin<1, time=100; end                  % line 2
```

```
% Initialize parameters n=100;
% D=0.2; gamma=0.5;
n=200; D=0.2; gamma=0.5;                    % line 3

%Set initial values of u randomly
u=rand(n,n);  grad=u*0;                     % line 4

% Index for u(i,j) and the loop
I = 2:n-1; J = 2:n-1;                       % line 5

% Time stepping
for step=1:time,                            % line 6
% Laplace gradient of the equation          % line 7
 grad(I,J)= u(I,J-1)+u(I,J+1)+u(I-1,J)+u(I+1,J);
 u =(1-4*D)*u+D*grad+gamma*u.*(1-u);        % line 8
% Show results
 pcolor(u);  shading interp;                % line 9
% Coloring and colorbar
 colorbar; colormap jet;                    % line 10
 drawnow;                                   % line 11
end                                         % line 12

% plot as a surface
surf(u);                                    % line 13
shading interp;                             % line 14
view([-25 70]);                             % line 15
% ------------- End of Pragram -----------------
```

If you use this program to do the simulations, you will see that the pattern emerges naturally from the initially random background. Once the pattern is formed, it evolves gradually with time, but the characteristics such as the shape and structure of the patterns do not change much with time. In this sense, one can see beautiful and stable patterns.

## 18.3 Reaction-Diffusion System

The pattern formation in the previous section arises naturally from a single equation of nonlinear reaction-diffusion type. In many applications, we often have to simulate a system of nonlinear reaction-diffusion equations, and the variables are coupled in a complicated manner.

The pattern formation in the previous section comes from the instability of the nonlinear reaction diffusion system. In order to show this, let us use the following mathematical model for enzyme inhibition and cooperativity.

For example, the following system consists of two nonlinear equations

$$\frac{\partial u}{\partial t} = D_u(\frac{\partial^2 u}{\partial x^2} + \frac{\partial^2 u}{\partial y^2}) + f(u, v), \tag{18.13}$$

$$\frac{\partial v}{\partial t} = D_v(\frac{\partial^2 v}{\partial x^2} + \frac{\partial^2 v}{\partial y^2}) + g(u, v), \tag{18.14}$$

and

$$f(u, v) = au(1 - u) - \frac{bu - cv}{1 + u + v}, \tag{18.15}$$

$$g(u, v) = -\frac{vd}{1 + u + v}, \tag{18.16}$$

where $D_u$ and $D_v$ are diffusion coefficients, while $a, b, c, d$ are all constants. This reaction diffusion system may have instability if certain conditions are met.

The steady state solutions are obtained from $f(u_0, v_0) = 0$ and $g(u_0, v_0) = 0$. They are

$$u_0 = \frac{b}{2a}[\sqrt{1 + 4\frac{a^2}{b^2}} - 1], \qquad v_0 = 0. \tag{18.17}$$

Let $\psi = (u - u_0, v - v_0)$ be the small perturbation, then $\psi$ satisfies

$$\frac{\partial \psi}{\partial t} = D\nabla^2\psi + M\psi, \tag{18.18}$$

where

$$D = \begin{pmatrix} D_u & 0 \\ 0 & D_v \end{pmatrix}, \tag{18.19}$$

and

$$M = \frac{1}{(1+u_0)} \begin{pmatrix} -(2au_0+b) & a(1-u_0)+c \\ 0 & d \end{pmatrix}. \tag{18.20}$$

Writing $\psi$ in the form of

$$\psi = \sum e^{\lambda t} \psi_k, \tag{18.21}$$

where the summation is over all the wavenumbers $k$, we have

$$|M - \lambda I - Dk^2| = 0, \tag{18.22}$$

where I is a $2 \times 2$ unity matrix. This eigenvalue equation has two roots. Since $\Re(\lambda) > 0$ implies that instability, this requires that

$$\frac{D_v}{D_u} < \frac{d}{(2au_0+b)}. \tag{18.23}$$

The range of unstable wavenumbers between the two roots of $k^2$ at the bifurcation point is given by

$$k_{\pm}^2 = \frac{dD_u - D_v(2au_0+b)}{2D_u D_v(1+u_0)} [1 \pm \sqrt{1 + 4D_u D_v \delta}], \tag{18.24}$$

with

$$\delta = \frac{(2au_0+b)}{[dD_u - D_v(2au_0+b)]^2}. \tag{18.25}$$

If the unstable criteria are satisfied, any small random perturbation can generate complex patterns.

Similar to the nonlinear KPP equation (18.12), beautiful patterns also arise naturally in the following nonlinear system

$$\frac{\partial u}{\partial t} = D_a\Big(\frac{\partial^2 u}{\partial x^2} + \frac{\partial^2 u}{\partial y^2}\Big) + \gamma \tilde{f}(u, v), \tag{18.26}$$

$$\frac{\partial v}{\partial t} = D_b\Big(\frac{\partial^2 v}{\partial x^2} + \frac{\partial^2 v}{\partial y^2}\Big) + \beta \tilde{g}(u, v), \tag{18.27}$$

and

$$\tilde{f}(u, v) = u(1 - u), \qquad \tilde{g}(u, v) = u - u^2 v, \tag{18.28}$$

for the values $D_a = 0.2$, $D_b = 0.1$, $\gamma = 0.5$ and $\beta = 0.2$. With different functions $\tilde{f}(u, v)$ and $\tilde{g}(u, v)$, these equations can be used to simulate the pattern formation in a wide range of applications where nonlinear reaction-diffusion equations are concerned.

# Chapter 19

# Probability and Statistics

All the mathematical models and differential equations we have discussed so far are deterministic systems in the sense that given accurate initial and boundary conditions, the solutions of the system can be determined (the only exception is the chaotic system to a certain degree). There is no intrinsic randomness in the differential equations. In reality, randomness occurs everywhere, and not all models are deterministic. In fact, it is necessary to use stochastic models and sometimes the only sensible models are stochastic descriptions. In these cases, we have to deal with probability and statistics.

## 19.1 Probability

### 19.1.1 Randomness and Probability

Randomness such as roulette-rolling and noise arises from the lack of information, or incomplete knowledge of reality. It can also come from the intrinsic complexity, diversity and perturbations of the system. The theory of probability is mainly the studies of random phenomena so as to find non-random regularity.

For an experiment or trial such as rolling dices whose outcome depends on chance, the sample space $\Omega$ of the experiment is the set of all possible outcomes. The sample space can be either finite or infinite. For example, rolling a six-sided die will have six different outcomes, thus the sample space is $\Omega = \{1, 2, 3, 4, 5, 6\}$. The elements of a sample space are the outcomes, and each subset of a sample space is called an event. For example, the event $S = \{2, 4, 6\}$ is a subset of $\Omega$. In a sample space $\Omega$, the outcomes of an experiment is represented as numbers (1 for heads and 0 for tails for tossing coins). A real-valued variable that is defined for all the possible outcomes is referred to as a random variable, which is a function that associates a unique numerical value with every outcome of an experiment, and its actual value varies from trial to trial as the experiment is repeated. The values of a random variable can be discrete (such as 1 to 6 in rolling a single die) or continuous (such as the level of noise). If a random variable only takes discrete values, it is called a discrete random variable. If its values are continuous, then it is called a continuous random variable.

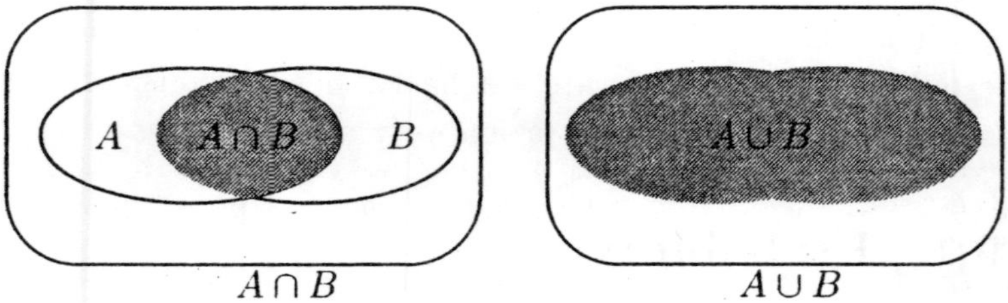


Figure 19.1: Venn Diagrams: $A \cap B$ and $A \cup B$.

Two events $A$ and $B$ can have various relationships and these can be represented by Venn diagrams as shown in Figure 19.1. The intersection $A \cap B$ of two events means the outcome of the random experiments belongs to both $A$ and $B$, and it is the case of '$A$ AND $B$'. If no event or outcome belongs to the intersection, that is $A \cap B = \emptyset$, we say these two events are mutually exclusive or disjoint.

The union $A \cup B$ denotes the outcome belongs to either $A$ or $B$ or both, and this means the case of '$A$ OR $B$'. The complement $\bar{A} = \Omega - A$ (or not $A$) of the event $A$ is the set of outcomes that do not belong to $A$ but in the sample space $\Omega$ (see Figure 19.2). The $A - B$ means the outcomes in $A$ only.

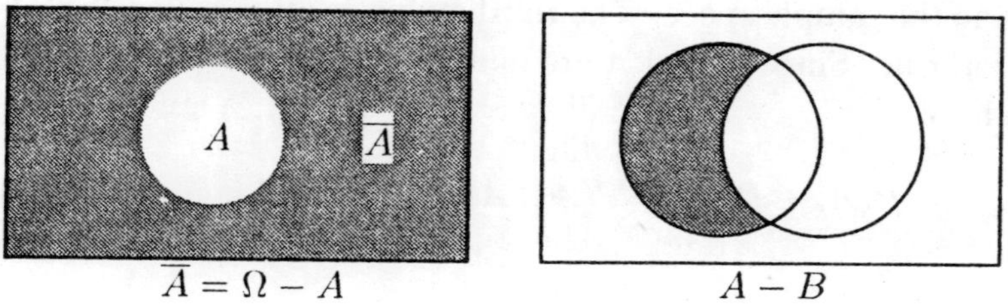


Figure 19.2: Venn Diagrams: $\bar{A} = \Omega - A$ and $A - B$.

Probability $P$ is a number or an expected frequency assigned to an event $A$ that indicates how likely the event will occur when a random experiment is performed. This probability is often written as $P(A)$ to show that the probability $P$ is associated with event $A$. For a large number of fair trials, the probability can be calculated by

$$P(A) = \frac{N_A(\text{number of outcomes in the event A})}{N_\Omega(\text{total number of outcomes})}. \tag{19.1}$$

---

□ ***Example 19.1:*** *If you tossed a coin 1000 times, the head (H) occurs 511 times and the tail (T) occurs 489 times. The probability $P(H)$ and $P(T)$ are*

$$P(H) = \frac{511}{100} = 0.511,$$

*and*

$$P(T) = \frac{489}{1000} = 0.489.$$

□

---

There are three axioms of probability, and they are:

Axiom I : $0 \leq P(A) \leq 1$.

Axiom II : $P(\Omega) = 1$.

Axiom III : $P(A \cup B) = P(A) + P(B)$, if $A \cap B = \emptyset$.

The first axiom says that the probability is a number between 0 and 1 inclusive. $P(A) = 0$ corresponds to impossibility while $P(A) = 1$ corresponds to absolute certainty. The second axiom simply means that an event must occur somewhere inside the sample space. The third axiom is often called the addition rule. Since $A$ and $\bar{A}$ are mutually exclusive ($A \cap \bar{A} = \emptyset$), we have

$$P(A) + P(\bar{A}) = P(A \cup \bar{A}) = P(\Omega) = 1, \tag{19.2}$$

or

$$P(A) = 1 - P(\bar{A}), \tag{19.3}$$

which is usually called the NOT rule. The third axiom can be further generalized to any two events $A$ and $B$

$$P(A \cup B) = P(A) + P(B) - P(A \cap B). \tag{19.4}$$

In a special case when events $A_i (i = 1, 2, ..., n)$ exhaust the whole sample space such that $A = \cup_{i=1}^{n} A_i = A_1 \cup A_i \cup ... \cup A_n = \Omega$ and $A_i \cap A_j = \emptyset (i \neq j)$,

$$P(A \cap B) = \sum_{i=1}^{n} P(A_i \cap B). \tag{19.5}$$

Since $\Omega \cap B = B$, we also get

$$P(\Omega \cap B) = P(B) = \sum_{i=1}^{n} P(A_i \cap B), \tag{19.6}$$

which are the useful properties of the total probability.

For example, if you randomly draw a card from a standard pack of 52 cards, what is the probability of it being a red king or a diamond with a face value being a prime number (if its face value is counted from 1 to 13). The prime numbers are 2, 3, 5, 7, 11, 13, therefore they are 6 cards that forms the primes.

The possibility of event (A) of drawing a red king is $P(A) = \frac{2}{52} = \frac{1}{26}$. The probability of event (B) of drawing a prime number is $P(B) = \frac{6}{52} = \frac{3}{26}$. As a diamond king (13) is also a prime, this means $P(A \cap B) = \frac{1}{52}$. Therefore, the probability

$$P(A \cup B) = P(A) + P(B) - P(A \cap B)$$
$$= \frac{1}{26} + \frac{3}{26} - \frac{1}{52} = \frac{7}{52}.$$

Two events $A$ and $B$ are independent if the events has no influence on each other. That is to say, the occurrence of one of the events does not provide any information about whether or the other event will occur. In this case, the probability of both occuring is equal to the product of the probabilities of the two individual events $P(A)$ and $P(B)$

$$P(A \cap B) = P(A) * P(B). \tag{19.7}$$

This can be easily extended to $n$ mutually independent events $A_i (i = 1, 2, ..., n)$. The probability of all these events happening is

$$P(\sum_{i=1}^{n} \cap A_i) = \Pi_{i=1}^{n} P(A_i) = P(A_1)P(A_2) \cdots P(A_n). \tag{19.8}$$

---

□ ***Example 19.2:*** *The probability of drawing a king from a pack of cards (Event A), and showing an even number of rolling a six-sided die (event B) is $P(A \cap B)$. We know $P(A) = 4/52 = 1/13$, and $P(B) = 3/6 = 1/2$. Since these two events are independent, the probability that both events occur is*

$$P(A \cap B) = P(A)P(B) = \frac{1}{13} \cdot \frac{1}{2} = \frac{1}{26}.$$

□

---

If the two events are not independent, then one may affect the other event, in this case, we are dealing with the conditional probability which will be discussed later in the next section.

In calculating the probabilities, it is useful to know the possible combinations and permutations of certain objects. Suppose you have 5 pairs of shoes, 4 pairs of trousers, 7 shirts and 2 hats. This is equivalent to the lineup problem from your feet to your head. In this case, as the event of selecting each thing to wear is in a similar manner of putting it into slots in successive stages, the total number of all possible ways is simply the multiplication of all the possible choices for each stages. All possible outfits you can wear form a permutation problem, and the total number is $5 \times 4 \times 7 \times 5 = 700$.

In order to line 5 objects marked $A, B, C, D, E$, in the first place, there are 5 possible choices, the second place has only 4 options, the third place 3 choices, the fourth place has 2 choices, and there is only one left for the last place. Thus the number of all possible permutations is $5 \times 4 \times 3 \times 2 \times 1 = 5!$. Following this line of reasoning, $n$ objects can in general be permutated in $n!$ ways.

Suppose there are $n = 20$ students in a class (named $S_1$, $S_2$, ...., $S_{20}$), we want to select 5 students at random to form a 5-student team to work on a project. This is different from the lineup problem because once you have selected any five students (say) $S_1$, $S_7$, $S_{10}$, $S_{15}$, $S_{19}$, it does not matter what order you selected them, the final formed team is the same. There are 5! permutations within the same team. Order does not count in this case. This is a combination problem (also called a committee problem). As before, there are 5 places to line up the students, and the total number of all permutations for selecting 5 students is $20 * 19 * 18 * 17 * 16$. Therefore, the total number of combinations (of selecting 5 students) is

$$^{20}C_5 = \frac{20 * 19 * 18 * 17 * 15}{5!} = \frac{20!}{5!15!} = 15504. \qquad (19.9)$$

In general, the total number of all possible combinations of selecting $k$ objects from $n$ is

$$^nC_k \equiv \begin{pmatrix} n \\ k \end{pmatrix} \equiv \frac{n!}{k!(n-k)!}. \qquad (19.10)$$

The consistency requires $0! = 1$.

---

□ ***Example 19.3:*** *A club of 5 members is chosen at random from 8 female students, 10 male students, and 7 teachers. What is the probability of the club consisting of 2 female students, 2 male students, and 1 teacher? The total number of clubs is* $^{25}C_5$. *If two female students are selected, we have* $^8C_2$. *Similarly,* $^{10}C_2$ *for selecting 2 male students, and* $^7C_1$ *for selecting one teacher. Therefore, the total number N of forming the 5-member club is*

$$N = \frac{^8C_2{}^{10}C_2{}^7C_1}{^{25}C_5} = \frac{42}{253} \approx 0.166$$

□

---

There is an interesting 'birthday paradox' which is related to this context. The birthday paradox was first proposed in 1939 by Richard von Mises, which states that what is the probability of two people having the same birthday in a group of $n$ people. For a group of 367 people, it is certain that there must be two people having the same birthday as there are only 365 (or 366 if someone was born in a leap year) possible birthdays. Ignoring 29 February and the year of birth and assuming that the birthdays are evenly distributed throughout the year, we only have 365 different birthdays (days and months only). If the event $A$ denotes that all the $n$ people will have different birthdays (no birthday matching), the first person can have any date as his or her birthday, 365/365. The second person must be in other 364 dates, which is 364/365, and the $k$th person has $(365 - k + 1)/365$. Therefore, the probability of no two people having the same birthday is

$$P(A, n) = \frac{365}{365} \times \frac{364}{365} \times ... \times \frac{(365 - n + 1)}{365}$$

$$= \frac{365 * (364) * ... * (365 - n + 1)}{365^n} = \frac{365!}{(365 - n)! 365^n}. \quad (19.11)$$

Now the probability of two people with the same birthday is

$$P(\bar{A}, n) = 1 - P(A, n) = 1 - \frac{365!}{(365 - n)! 365^n}. \quad (19.12)$$

The factorial 365! is a large number, but you do not have to deal with such large numbers. You can use a simple calculator to estimate it. For five people, the probability of two people with the same birthday is

$$P(\bar{A}, 5) = 1 - \frac{365 * 364 * 363 * 362 * 361}{365^5} \approx 0.027, \quad (19.13)$$

which seems insignificant. However, the interesting thing is that for $n = 23$, the probability becomes

$$P(\bar{A}, 23) = 1 - \frac{365!}{(365 - 23)!365^{23}} \approx 0.507. \quad (19.14)$$

This means that you have slightly more than a 50-50 chance of finding two people sharing the same birthday. If you increase $n$, you get $P(\bar{A}, 30) \approx 0.706$ for $n = 30$, $P(\bar{A}, 40) \approx 0.891$ for $n = 40$, and $P(\bar{A}, 50) \approx 0.970$ and $P(\bar{A}, 70) \approx 0.9992$ (almost certainty) for $n = 70$.

It is worth noting that there is some difference in combinations when the member drawn is placed back or not. Suppose there are 10 red balls and 10 white balls in bag. If we draw a ball (say a red, event A) from the bag and then put it back (with replacement), then we draw another ball (event B). $P(A) = 1/20$ and $P(B) = 1/20$. The probability of getting two red balls are $P(A \cap B) = P(A) * P(B) = 1/400$. We call this case I.

For a second case (Case II), if we do not put it back after we have drawn the first ball (without replacement), then the probability of event B is now different $P(B) = 1/19$ as there is now only 19 balls in the bag. The probability of getting two red balls now becomes $P(A \cap B) = \frac{1}{20} \times \frac{1}{19} = \frac{1}{380}$, which is different from $1/400$.

The reason here is that the two events are not independent in the case of no-replacement. If we use notation '$B|A$' which means that event B occurs given that event A has occurred, then we can use $P(B|A)$ to denote the probability of event B when there is no replacement in event A in the scenario described in Case II. Now $P(B)$ becomes $P(B|A)$. Hence, we have

$$P(A \cap B) = P(A)P(B|A), \tag{19.15}$$

which is often called the multiplication rule in probability theory. Similarly, we can get

$$P(A \cap B) = P(B)P(A|B). \tag{19.16}$$

This is essentially a conditional probability problem which forms the main topic of the next section.

### 19.1.2 Conditional Probability

In calculating the probabilities, we often assume that all possible outcomes of an experiment such as drawing a card are equally likely. Probabilities can change if additional information is known or some other event has already occurred and thus $P(B|A)$ denotes the probability that event B will occur given that event A has already occurred. The conditional probability can be calculated by

$$P(B|A) = \frac{P(B \cap A)}{P(A)}. \tag{19.17}$$

Conversely, we have

$$P(A|B)\frac{P(A \cap B)}{P(B)}. \tag{19.18}$$

Using equation (19.15), we can write the above formulation as

$$P(A|B) = \frac{P(A)P(B|A)}{P(B)}$$

$$= \frac{P(A)P(B|A)}{P(A)P(B|A) + P(\bar{A})P(B|\bar{A})}, \tag{19.19}$$

which is the Bayes' theorem. Here have used $\bar{A} \cup A = \Omega$ and $P(\bar{A}) = 1 - P(A)$.

As an example, we consider the drug test in sports. It is believed that the test is 99% accurate if athletes are taking drugs. For athletes not taking drugs, the positive test is only 0.5%. It is assumed that only one in 1000 athletes takes this kind of drug. Suppose an athlete is selected at random and the test shows positive for the drug. What is the probability that the athlete is really taking the drug? If event $A$ denotes an athlete is taking the drug, and $B$ denotes the event that the individual tests positive. Thus, $P(A) = 1/1000$, $P(B|A) = 0.99$ and $P(B|\bar{A}) = 0.005$. The probability that the athlete is actually taking the drug is

$$P(A|B) = \frac{P(A)P(B|A)}{P(A)P(B|A) + P(\bar{A})P(B|\bar{A})}$$

$$= \frac{0.001 * 0.99}{0.001 * 0.99 + 0.999 * 0.005} \approx 0.165. \tag{19.20}$$

This is surprisingly low in probability.

---

□ ***Example 19.4:*** *The classical problem of three cards consists of three cards: one blue card (B) is blue on both sides, one white card (W) is white on both sides, and one mixed card (M) is white on one side and blue on the other. If you draw one card at random from a bag and place it on a table, suppose that the side you can see is blue, what is the probability of other side is also blue? This a conditional probability problem. There are 3 blue faces and 3 white faces, thus the total probability of showing a blue face (F) is $P(BF) = 1/2$, and probability of pull the blue-blue card out is $P(BB) = 1/3$, while the probability of showing a blue face is $P(BF|BB) = 1$ if the pulled card is blue-blue one. Then, the probability of other side being also blue is*

$$P(BB|BF) = \frac{P(BF|BB)P(BB)}{P(BF)} = \frac{1 \times \frac{1}{3}}{\frac{1}{2}} = \frac{2}{3}. \tag{19.21}$$

*Most people will intuitively guess that the probability is $\frac{1}{2}$, which is not correct.* □

---

Another related problem is the so-called Monty Hall problem (or three door problem) in a classical game show. Suppose that you are given the choice of three doors. There is an expensive car behind one door, behind other two doors are goats. If you choose one door (say, A) at random, then the host opens one of the other door (say, B), which he knows there is a goat behind it, to reveal a goat. Now you have a choice either to stick with your original choice or swap with the other door (say, C). What is your best strategy based on probability? Initially, the prize car behind any door (Y) has a priori probability $P(any) = 1/3$, so your initial choice $P(A) = 1/3$. As the host knows where the prize is, if the car is behind $A$, the host will open B or C so 1/2 each. If the car is behind $B$, the host will never open $B$, and if the car is behind C, the host will surely open $B$. Mathematically, this gives $P(OpenB|A) = 1/2$, $P(OpenB|B) = 0$, and $P(OpenB|C) = 1$.

So the total probability of opening door $B$ is

$$P(OpenB) = P(A)P(OpenB|A) + P(B)P(OpenB|B)$$

$$+P(C)P(OpenB|C) = \frac{1}{3} \times \frac{1}{2} + \frac{1}{3} \times 0 + \frac{1}{3} \times 1 = \frac{1}{2}. \quad (19.22)$$

Now the probability of the car behind door $C$ is

$$P(C|OpenB) = \frac{P(OpenB|C)P(C)}{P(OpenB)} = \frac{1 \times \frac{1}{3}}{\frac{1}{2}} = \frac{2}{3}, \quad (19.23)$$

which is greater than your initial choice 1/3. Therefore, the best strategy is to switch your choice. This game has other variations such as the three envelope problem and others, but the analysis and strategy are the same.

### 19.1.3 Random Variables and Moments

#### Random Variables

For a discrete random variable $X$ with distinct values such as the number of cars passing through a junction, each value $x_i$ may occur with a certain probability $p(x_i)$. In other words, the probability varies with the random variable. A probability function $p(x_i)$ is a function that defines probabilities to all the discrete values $x_i$ of the random variable $X$. As an event must occur inside a sample space, the requirement that all the probabilities must be summed to one leads to

$$\sum_{i=1}^{n} p(x_i) = 1. \tag{19.24}$$

The cumulative probability function of $X$ is defined by

$$P(X \leq x) = \sum_{x_i < x} p(x_i). \tag{19.25}$$

For a continuous random variable $X$ that takes a continuous range of values (such as the level of noise), its distribution is continuous and the probability density function $p(x)$ is defined for a range of values $x \in [a, b]$ for given limits $a$ and $b$ [or even over the whole real axis $x \in (-\infty, \infty)$]. In this case, we always use the interval $(x, x + dx]$ so that $p(x)$ is the probability that the random variable $X$ takes the value $x < X \leq x + dx$ is

$$\Phi(x) = P(x < X \leq x + dx) = p(x)dx. \tag{19.26}$$

As all the probabilities of the distribution shall be added to unity, we have

$$\int_a^b p(x)dx = 1. \tag{19.27}$$

The cumulative probability function becomes

$$\Phi(x) = P(X \leq x) = \int_a^x p(x)dx, \tag{19.28}$$

which is the definite integral of the probability density function between the lower limit $a$ up to the present value $X = x$.

### Mean and Variance

Two main measures for a random variable $X$ with a given probability distribution $p(x)$ are its mean and variance. The mean

$\mu$ or the expectation value of $E[X]$ is defined by

$$\mu \equiv E[X] \equiv <X> = \int x p(x) dx, \tag{19.29}$$

for a continuous distribution and the integration is within the integration limits. If the random variable is discrete, then the integration becomes the summation

$$E[X] = \sum_i x_i p(x_i). \tag{19.30}$$

The variance $\text{var}[X] = \sigma^2$ is the expectation value of the deviation squared $(X-\mu)^2$. That is

$$\sigma^2 \equiv \text{var}[X] = E[(X-\mu)^2] = \int (x-\mu)^2 p(x) dx. \tag{19.31}$$

The square root of the variance $\sigma = \sqrt{\text{var}[X]}$ is called the standard deviation, which is simply $\sigma$.

This simply becomes a sum

$$\sigma^2 = \sum_i (x-\mu)^2 p(x_i), \tag{19.32}$$

for a discrete distribution. In addition, any other formulas for a continuous distribution can be converted to their counterpart for a discrete distribution if the integration is replaced by the sum. Therefore, we will mainly focus on the continuous distribution in the rest of the section.

Other frequently used measures are the mode and median. The mode of a distribution is defined by the value at which the probability density function $p(x)$ is maximum. For an even number of data sets, the mode may have two values. The median $m$ of a distribution corresponds to the value at which the cumulative probability function $\Phi(m) = 1/2$. The upper and lower quartiles $Q_U$ and $Q_L$ are defined by $\Phi(Q_U) = 3/4$ and $\Phi(Q_L) = 1/4$.

### Moments and Moment Generating Functions

In fact, the mean is essentially the first moment if we define the $k$th moment of a random variable $X$ by

$$E[X^k] \equiv \mu_k = \int x^k p(x) dx, \qquad k = 1, 2, ..., N. \tag{19.33}$$

Similarly, the $k$th central moment is defined by

$$E[(X - \mu)^k] \equiv \nu_k = \int (x - \mu)^k p(x) dx, \qquad k = 1, 2, ..., N. \tag{19.34}$$

Obviously, the variance is the second central moment. From these definitions, it is straightforward to prove

$$E[\alpha x + \beta] = \alpha E[X] + \beta, \qquad E[X^2] = \mu^2 + \sigma^2, \tag{19.35}$$

and

$$\text{var}[\alpha x + \beta] = \alpha^2 \text{var}[X]. \tag{19.36}$$

where $\alpha$ and $\beta$ are constants.

Most probability functions can be expressed in terms of moments and moment generating functions. The moment generating function is defined by

$$G_X(\nu) \equiv E[e^{\nu X}] = \int e^{\nu x} p(x) dx, \tag{19.37}$$

where $\nu \in \mathcal{R}$ is a real parameter. By expanding $\exp[\nu x]$ into power series and using the definition of various moments, it is straightforward to verify that

$$E[X^k] = \frac{d^k G_X(\nu)}{d\nu^k}\Big|_{\nu=0}, \tag{19.38}$$

and

$$\sigma^2 = \frac{d^2 G_X(0)}{d\nu^2} - [\frac{dG_X(0)}{d\nu}]^2. \tag{19.39}$$

### 19.1.4 Binomial and Poisson Distributions

#### Binomial Distribution

A discrete random variable is said to follow the binomial distribution $B(n,p)$ if its probability distribution is given by

$$B(n,p) = {}^nC_x p^x (1-p)^{n-x}, \qquad {}^nC_x = \frac{n!}{x!(n-x)!}, \tag{19.40}$$

where $x = 0, 1, 2, ..., n$ are the values that the random variable $X$ may take, $n$ is the number of trials. There are only two possible outcomes: success or failure. $p$ is the probability of a so-called 'success' of the outcome. Subsequently, the probability of the failure of a trial is $q = 1 - p$. Therefore, $B(n,p)$ represents the probability of $x$ successes and $n - x$ failures in $n$ trials. The coefficients come from the coefficients of the binomial expansions

$$(p+q)^n = \sum_{x=0}^{n} {}^nC_x p^x q^{n-x} = 1, \tag{19.41}$$

which is exactly the requirement that all the probabilities should be summed to unity.

---

□ ***Example 19.5:*** *Tossing a coin 10 times, the probability of getting 7 heads is $B(n, 1/2)$. Since $p = 1/2$ and $x = 7$, then we have*

$${}^{10}C_7 (\frac{1}{2})^7 (\frac{1}{2})^3 = \frac{15}{128} \approx 0.117.$$

□

---

It is straightforward to prove that $\mu = E[X] = np$ and $\sigma^2 = npq = np(1-p)$ for a binomial distribution.

Another related distribution is the geometric distribution whose probability function is defined by

$$P(X = n) = pq^{n-1} = p(1-p)^{n-1}, \tag{19.42}$$

where $n \geq 1$. This distribution is used to calculate the first success, thus the first $n - 1$ trials must be in failure if $n$ trials are needed to observe the first success. The mean and variance of this distribution are $\mu = 1/p$ and $\sigma^2 = (1-p)/p^2$.

## Poisson Distribution

The Poisson distribution can be thought as the limit of the binomial distribution when the number of trial is very large $n \to \infty$ and the probability $p \to 0$ (small probability) with the constraint that $\lambda = np$ is finite. For this reason, it is often called the distribution for small-probability events. Typically, it is concerned with the number of events that occur in a certain time interval (e.g., number of telephone calls in an hour) or spatial area. The Poisson distribution is

$$P(X = x) = \frac{\lambda^x e^{-\lambda}}{x!}, \qquad \lambda > 0, \tag{19.43}$$

where $x = 0, 1, 2, ..., n$ and $\lambda$ is the mean of the distribution. Using the definition of mean and variance, it is straightforward to prove that $\mu = \lambda$ and $\sigma^2 = \lambda$ for the Poisson distribution. The parameter $\lambda$ is the location of the peak as shown in Figure 19.3.

---

□ ***Example 19.6:*** *If you receive 3 calls per hour on your mobile phone on the average. If you do not switch your phone off, what is the probability that it begins to sound (one call is enough) during any one-hour class? We know that $\lambda = 3$. The probability of no phone call at all is*

$$P(X = 0) = \frac{3^0 e^{-3}}{0!} \approx 0.0498.$$

*Thus, the probability of sounding is $P(X > 0) \approx 1 - 0.0498 \approx 0.95$. In fact, the probability of receiving one call is*

$$P(X = 1) = \frac{3^1 e^{-3}}{1!} \approx 0.149,$$

*and the probability of receiving two calls is*

$$P(X = 2) = \frac{3^2 e^{-3}}{2!} \approx 0.224.$$

□

---

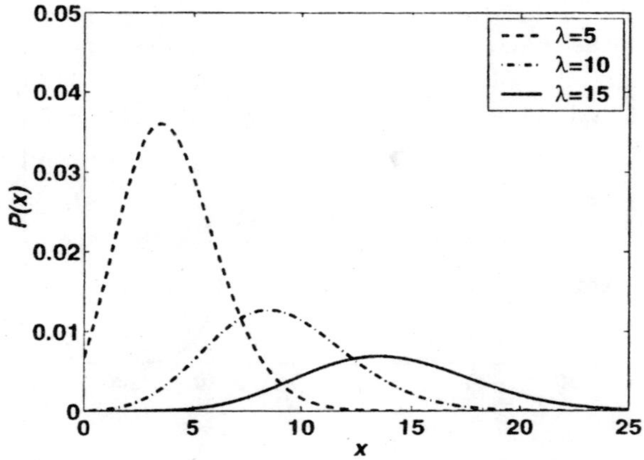


Figure 19.3: Poisson distributions for different values of $\lambda = 5, 10, 15$.

The moment generating function for the Poisson distribution is given by

$$G_X(\nu) = \sum_{x=0}^{\infty} \frac{e^{\nu x}\lambda^x e^{-\lambda}}{x!} = \exp[\lambda(e^\nu - 1)]. \tag{19.44}$$

### 19.1.5 Gaussian Distribution

The Gaussian distribution or normal distribution is the most important continuous distribution in probability and it has a wide range of applications. For a continuous random variable $X$, the probability density function (PDF) of a Gaussian distribution is given by

$$p(x) = \frac{1}{\sigma\sqrt{2\pi}} e^{-\frac{(x-\mu)^2}{2\sigma^2}}. \tag{19.45}$$

where $\sigma^2 = \text{var}[X]$ is the variance and $\mu = E[X]$ is the mean of the Gaussian distribution. From the Gaussian integral, it is easy to verify that

$$\int_{\infty}^{\infty} p(x)dx = 1, \tag{19.46}$$

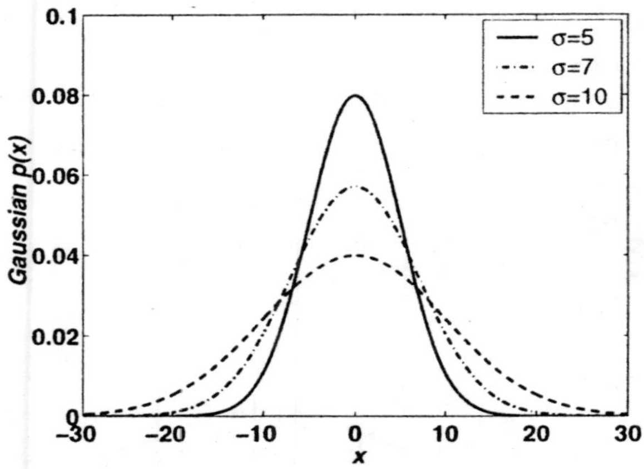


Figure 19.4: Gaussian distributions for $\sigma = 5, 7, 10$.

and this is exactly the reason that the factor $1/\sqrt{2\pi}$ comes from the normalization of the all probabilities. The probability function reaches a peak at $x = \mu$ and the variance $\sigma^2$ controls the width of the peak (see Figure 19.4).

The cumulative probability function (CPF) for a normal distribution is the integral of $p(x)$, which is defined by

$$\Phi(x) = P(X < x) = \frac{1}{\sqrt{2\pi\sigma^2}} \int_{-\infty}^{x} e^{-\frac{(\zeta-\mu)^2}{2\sigma^2}} d\zeta. \qquad (19.47)$$

Using the error function defined by Chapter 1, we can write it as

$$\Phi(x) = \frac{1}{\sqrt{2}}[1 + \mathrm{erf}(\frac{x-\mu}{\sqrt{2}\sigma})]. \qquad (19.48)$$

The moment generating function for the Gaussian distribution is given by

$$G_X(\nu) = e^{\mu\nu + \frac{1}{2}(\sigma\nu)^2}. \qquad (19.49)$$

The Gaussian distribution can be considered as the limit of the Poisson distribution when $\lambda \gg 1$. Using the Sterling's approximation $x! \sim \sqrt{2\pi x}(x/e)^x$ for $x \gg 1$, and setting $\mu = \lambda$ and $\sigma^2 = \lambda$, it can be verified that the Poisson distribution can be written as a Gaussian distribution

$$P(x) \approx \frac{1}{\sqrt{2\pi\lambda}} e^{-\frac{(x-\mu)^2}{2\lambda}}, \tag{19.50}$$

where $\mu = \lambda$. In statistical applications, the normal distribution is often written as $N(\mu, \sigma)$ to emphasize that the probability density function depends on two parameters $\mu$ and $\sigma$.

The standard normal distribution is a normal distribution $N(\mu, \sigma)$ with a mean of $\mu = 0$ and standard deviation $\sigma = 1$, that is $N(0, 1)$ This is useful to normalize or standardize data for statistical analysis. If we define a normalized variable

$$\xi = \frac{x - \mu}{\sigma}, \tag{19.51}$$

it is equivalent to give a score so as to place the data above or below the mean in the unit of standard deviation. In terms of the area under the probability density function, $\xi$ sorts where the data falls. It is worth pointing out that some books define $z = \xi = (x - \mu)/\sigma$ in this case, and call the standard normal distribution as the Z distribution.

Table 19.1: Function $\phi$ defined by equation (19.53).

| $\xi$ | $\phi(\xi)$ | $\xi$ | $\phi$ |
|---|---|---|---|
| 0.0 | 0.500 | 1.0 | 0.841 |
| 0.1 | 0.540 | 1.1 | 0.864 |
| 0.2 | 0.579 | 1.2 | 0.885 |
| 0.3 | 0.618 | 1.3 | 0.903 |
| 0.4 | 0.655 | 1.4 | 0.919 |
| 0.5 | 0.692 | 1.5 | 0.933 |
| 0.6 | 0.726 | 1.6 | 0.945 |
| 0.7 | 0.758 | 1.7 | 0.955 |
| 0.8 | 0.788 | 1.8 | 0.964 |
| 0.9 | 0.816 | 1.9 | 0.971 |

Now the probability density function of standard normal distribution becomes

$$p(x) = \frac{1}{\sqrt{2\pi}} e^{-\frac{\xi^2}{2}}. \tag{19.52}$$

Its cumulative probability function is

$$\phi(\xi) = \frac{1}{\sqrt{2\pi}} \int_{-\infty}^{\xi} e^{-\frac{\xi^2}{2}} d\xi = \frac{1}{2}[1 + \text{erf}(\frac{\xi}{\sqrt{2}})]. \tag{19.53}$$

As the calculations of $\phi$ and the error function involve the numerical integrations, it is usual practice to tabulate $\phi$ in a table (see Table 19.1) so that you do not have to calculate their values each time you use it.

### 19.1.6 Other Distributions

There are a dozen of other important distributions such as the exponential distribution, log-normal distribution, uniform distribution and the $\chi^2$-distribution. The uniform distribution has a probability density function

$$p = \frac{1}{\beta - \alpha}, \qquad x = [\alpha, b], \tag{19.54}$$

whose mean is $E[X] = (\alpha + \beta)/2$ and variance is $\sigma^2 = (\beta - \alpha)^2/12$.

The exponential distribution has the following probability density function

$$f(x) = \lambda e^{-\lambda x} \quad (x > 0), \tag{19.55}$$

and $f(x) = 0$ for $x \leq 0$. Its mean and variance are

$$\mu = 1/\lambda, \qquad \sigma^2 = 1/\lambda^2. \tag{19.56}$$

The log-normal distribution has a probability density function

$$f(x) = \frac{1}{x\sqrt{2\pi\sigma^2}} \exp[-\frac{(\ln x - \mu)^2}{2\sigma^2}], \tag{19.57}$$

whose mean and variance are

$$E[X] = e^{\mu+\sigma^2/2}, \qquad \text{var}[X] = e^{\sigma^2+2\mu}(e^{\sigma^2} - 1). \tag{19.58}$$

The $\chi^2$-distribution, called chi-square or chi-squared distribution, is very useful in statistical inference and method of least squares. This distribution is for the quantity

$$\chi_n^2 = \sum_{i=1}^{n} (\frac{X_i - \mu_i}{\sigma_i})^2, \tag{19.59}$$

where the $n$-independent variables $X_i$ are normally distributed with means $\mu_i$ and variances $\sigma_i^2$. The probability density function for $\chi^2$-distribution is given by

$$p(x) = \frac{1}{2^{n/2}\Gamma(n/2)} x^{\frac{n}{2}-1} e^{-x/2}, \tag{19.60}$$

where $x \geq 0$, and $n$ is called the degree of freedom. Its cumulative distribution function is

$$\Phi(x) = \frac{\gamma(n/2, x/2)}{\Gamma(n/2)}, \tag{19.61}$$

where $\gamma(n/2, x/2$ is the incomplete gamma function. It can be verified that the mean of the distribution is $n$ and its variance is $2n$.

For other distributions, readers can refer to any books that are devoted to probability theory and statistical analysis.

### 19.1.7 The Central Limit Theorem

The most important theorem in probability is the central limit theorem which concerns the large number of trials and explains why the normal distribution occurs so widely. This theorem is as follows: Let $X_i (i = 1, 2, ..., n)$ be $n$ independent random variables, each of which is defined by a probability density function $p_i(x)$ with a corresponding mean $\mu_i$ and a variance $\sigma_i^2$. The sum of all these random variables

$$\Theta = \sum_{i=1}^{n} X_i = X_1 + X_2 + ... + X_n, \tag{19.62}$$

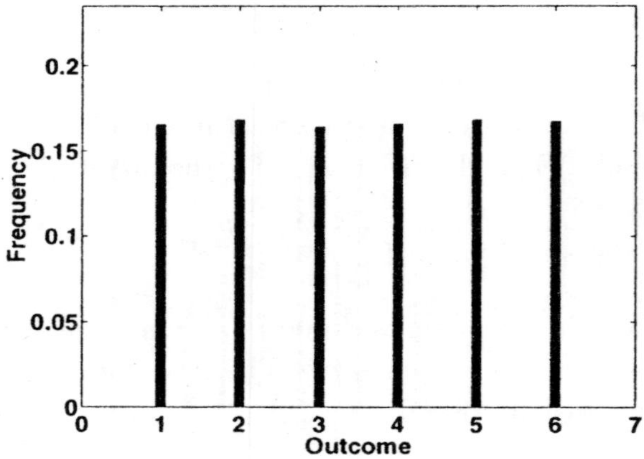


Figure 19.5: A uniform distribution.

is also a random variable whose distribution approaches the Gaussian distribution as $n \to \infty$. Its mean $E[\Theta]$ and variance $\text{var}[\Theta]$ are given by

$$E[\Theta] = \sum_{i=1}^{n} E[X_i] = \sum_{i=1}^{n} \mu_i, \tag{19.63}$$

and

$$\text{var}[\Theta] = \sum_{i=1}^{n} \text{var}[\Theta] = \sum_{i=1}^{n} \sigma_i^2. \tag{19.64}$$

The proof of this theorem is out of the scope of this book as it involves the moment generating functions, characteristics functions and other techniques. In engineering mathematics, we simply use these important results for statistical analysis.

In the special case when all the variables $X_i$ are described by the same probability density function with the same mean $\mu$ and variance $\sigma^2$, these results become

$$E[\Theta] = n\mu, \qquad \text{var}[\Theta] = n\sigma^2. \tag{19.65}$$

By defining a new variable

$$\xi_n = \frac{\Theta - n\mu}{\sigma\sqrt{n}}, \tag{19.66}$$

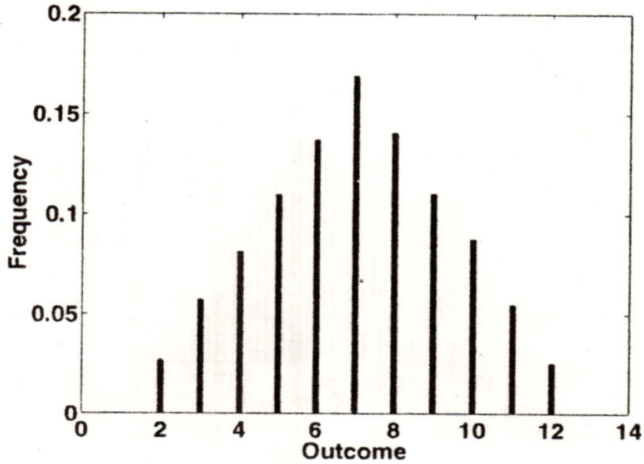


Figure 19.6: A bilinear distribution.

then the distribution of $\xi_n$ converges towards the standard normal distribution $N(0, 1)$ as $n \to \infty$.

Let us see what the theorem means for a simple experiment of rolling a few dice. For a fair six-sided die, each side will appear equally likely with a probability of $1/6 \approx 0.1667$, thus the probability function after rolling it 15000 times approaches a uniform distribution as shown in Figure 19.5.

If we now roll two independent dice 15000 times and count the sum (1-12) of the face values of both dice, then the sum obeys a bilinear distribution as shown in Figure 19.6. If we roll $n = 15$ independent dice, the sums of the face values vary from 1 to 90. After rolling the 15 dice 10,000 times, the distribution is shown in Figure 19.7 and it approaches to a normal distribution as $n \to \infty$.

## 19.2 Statistics

Statistics is the mathematics of data collection and interpretation, and the analysis and characterisation of numerical data by inference from sampling. Statistical methods involve reduc-

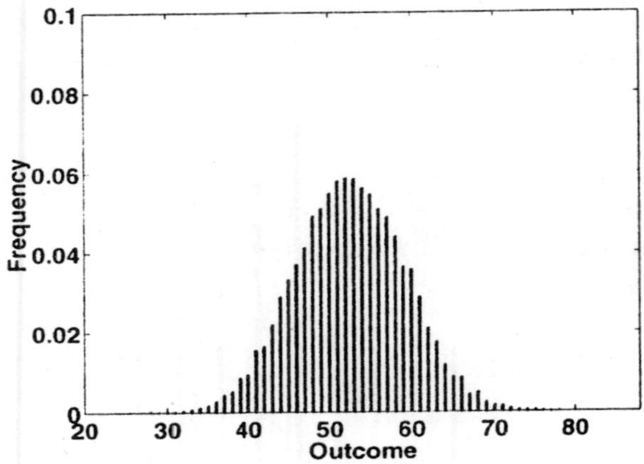


Figure 19.7: An approximate Gaussian distribution (the outcomes of the sum of face values in rolling 15 dice).

tion of data, estimates and significance tests, and relationship between two or more variables by analysis of variance, and the test of hypotheses.

### 19.2.1 Sample Mean and Variance

If a sample consists of $n$ independent observations $x_1, x_2, ..., x_n$ on a random variable $x$ such as the price of a cup of coffee, two important and commonly used parameters are sample mean and sample variance, which can easily be estimated from the sample. The sample mean is calculated by

$$\bar{x} \equiv <x> = \frac{1}{n}(x_1 + x_2 + ... + x_n) = \frac{1}{n}\sum_{i=1}^{n} x_i, \qquad (19.67)$$

which is essentially the arithmetic average of the values $x_i$.

Generally speaking, if $u$ is a linear combination of $n$ independent random variables $y_1, y_2, ..., y_n$ and each random variable $y_i$ has an individual mean $\mu_i$ and a corresponding variance

$\sigma_i^2$, we have the linear combination

$$u = \sum_{i=1}^{n} \alpha_i y_i = \alpha_1 y_1 + \alpha_2 y_2 + ... + \alpha_n y_n. \tag{19.68}$$

where the parameters $\alpha_i (i = 1, 2, ..., n)$ are the weighting coefficients. From the central limit theorem, we have the mean $\mu_u$ of the linear combination

$$\mu_u = E(u) = E(\sum_{i=1}^{n} \alpha_i y_i) = \sum_{i=1}^{n} \alpha E(y_i) = \sum \alpha_i \mu_i. \tag{19.69}$$

Then, the variance $\sigma_u^2$ of the combination is

$$\sigma_u^2 = E[(u - \mu_u)^2] = E\Big[\sum_{i=1}^{n} \alpha_i (y_i - \mu_i)^2\Big], \tag{19.70}$$

which can be expanded as

$$\sigma_u^2 = \sum_{i=1}^{n} \alpha_i^2 E[(y_i - \mu_i)^2]$$

$$+ \sum_{i,j=1:i\neq j}^{n} \alpha_i \alpha_j E[(y_i - \mu_i)(y_j - \mu_j)], \tag{19.71}$$

where $E[(y_i - \mu_i)^2] = \sigma_i^2$. Since $y_i$ and $y_j$ are independent, we have $E[(y_i - \mu_i)(y_j - \mu_j)] = E[(y_i - \mu_i)]E[(y_j - \mu_j)] = 0$. Therefore, we get

$$\sigma_u^2 = \sum_{i=1}^{n} \alpha_i^2 \sigma_i^2. \tag{19.72}$$

The sample mean defined in equation (19.67) can also be viewed as a linear combination of all the $x_i$ assuming each of which has the same mean $\mu_i = \mu$ and variance $\sigma_i^2 = \sigma^2$, and the same weighting coefficient $\alpha_i = 1/n$. Hence, the sample mean is an unbiased estimate of the sample due to the fact $\mu_{\bar{x}} = \sum_{i=1}^{n} \mu/n = \mu$. In this case, however, we have the variance

$$\sigma_{\bar{x}}^2 = \sum_{i=1}^{n} \frac{1}{n^2} \sigma^2 = \frac{\sigma^2}{n}. \tag{19.73}$$

which means the variance becomes smaller as the size $n$ of the sample increases by a factor of $1/n$.

The sample variance $S^2$ is defined by

$$S^2 = \frac{1}{n-1}\sum_{i=1}^{n}(x_i - \bar{x})^2. \tag{19.74}$$

It is worth pointing out that the factor is $1/(n-1)$ not $1/n$ because only $1/(n-1)$ will give the correct and unbiased estimate of the variance. From the probability theory in the earlier sections, we know that $E[x^2] = \mu^2 + \sigma^2$. The mean of the sample variance is

$$\mu_{S^2} = E[\frac{1}{n-1}\sum_{i=1}^{n}(x_i - \bar{x})^2] = \frac{1}{n-1}\sum_{i=1}^{n} E[(x_i^2 - n\bar{x}^2)]. \tag{19.75}$$

Using $E[\bar{x}^2] = \mu^2 + \sigma^2/n$, we get

$$\mu_{S^2} = \frac{1}{n-1}\sum_{i=1}^{n}\{E[x_i^2] - nE[\bar{x}^2]\}$$

$$= \frac{1}{n-1}\{n(\mu^2 + \sigma^2) - n(\mu^2 + \frac{\sigma^2}{n})\} = \sigma^2. \tag{19.76}$$

Obviously, if we use the factor $1/n$ instead of $1/(n-1)$, we would get $\mu_{S^2} = \frac{n-1}{n}\sigma^2 < \sigma^2$, which would underestimate the sample variance. The other way to think the factor $1/(n-1)$ is that we need at least one value to estimate the mean, we need at least 2 values to estimate the variance. Thus, for $n$ observations, only $n - 1$ different values of variance can be obtained to estimate the total sample variance.

### 19.2.2 Method of Least Squares

#### Maximum Likelihood

For a sample of $n$ values $x_1, x_2, ..., x_n$ of a random variable $X$ whose probability density function $p(x)$ depends on a set of $k$ parameters $\beta_1, ..., \beta_k$, the joint probability is then

$$\Phi(\beta_1, ..., \beta_k) = \Pi_{i=1}^{n} p(x_i, \beta_1, ..., \beta_k)$$

$$= p(x_1, \beta_1, ..., \beta_k)p(x_2, \beta_1, ..., \beta_k) \cdots p(x_n, \beta_1, ..., \beta_k). \quad (19.77)$$

The essence of the maximum likelihood is to maximize $\Phi$ by choosing the parameters $\beta_i$. As the sample can be considered as given values, the maximum likelihood requires that

$$\frac{\partial \Phi}{\partial \beta_i} = 0, \qquad (i = 1, 2, ..., k), \qquad (19.78)$$

whose solutions for $\beta_i$ are the maximum likelihood estimates.

### Linear Regression

For experiments and observations, we usually plot one variable such as pressure or price $y$ against another variable $x$ such as time or spatial coordinates. We try to present the data in a way so that we can see some trend in the data. For $n$ sets of data points $(x_i, y_i)$, the usual practice is to try to draw a straight line $y = a + bx$ so that it represents the major trend. Such a line is often called the regression line or the best fit line as shown in Figure 19.8.

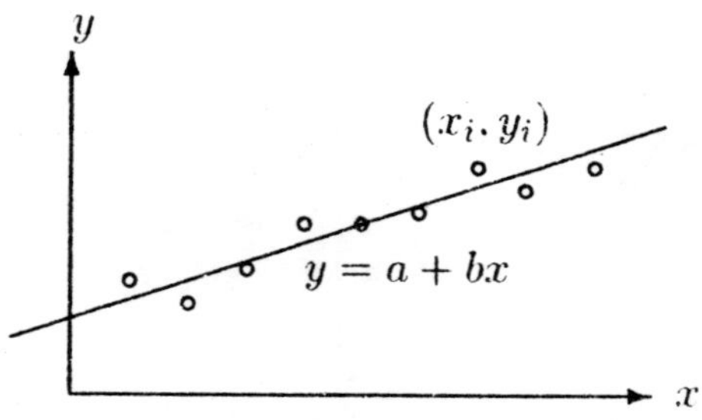


Figure 19.8: Least square and the best fit line.

The method of least squares is to try to determine the two parameters $a$ (intercept) and $b$ (slope) for the regression line from $n$ data points. Assuming that $x_i$ are known more precisely and $y_i$ values obey a normal distribution around the potentially best fit line with a variance $\sigma^2$. Hence, we have the probability

$$P = \Pi_{i=1}^n p(y_i) = A \exp\{-\frac{1}{2\sigma^2} \sum_{i=1}^n [y_i - f(x_i)]^2\}, \qquad (19.79)$$

where $A$ is a constant, and $f(x)$ is the function for the regression $[f(x) = a + bx$ for the linear regression$]$. It is worth pointing out that the exponent $\sum_{i=1}^{n}[y_i - f(x_i)]^2/\sigma_2$ is similar to the quantity $\chi_n^2$ defined in the $\chi^2$-distribution.

The essence of the method of least squares is to maximize the probability $P$ by choosing the appropriate $a$ and $b$. The maximization of $P$ is equivalent to the minimization of the exponent $\psi$

$$\psi = \sum_{i=1}^{n}[y_i - f(x_i)]^2. \tag{19.80}$$

We see that $\psi$ is the sum of the squares of the deviations $\epsilon_i^2 = (y_i - f(x_i))^2$ where $f(x_i) = a + bx_i$. The minimization means the least sum of the squares, thus the name of the method of least squares.

In order to minimize $\psi$ as a function of $a$ and $b$, its derivatives should be zero. That is

$$\frac{\partial\psi}{\partial a} = -2\sum_{i=1}^{n}[y - (a + bx_i)] = 0, \tag{19.81}$$

and

$$\frac{\partial\psi}{\partial b} = -2\sum_{i=1}^{n}x_i[y_i - (a + bx_i)] = 0. \tag{19.82}$$

By expanding these equations, we have

$$na + b\sum_{i=1}^{n}x_i = \sum_{i=1}^{n}y_i, \tag{19.83}$$

and

$$a\sum_{i=1}^{n}x_i + b\sum_{i=1}^{n}x_i^2 = \sum_{i=1}^{n}x_iy_i, \tag{19.84}$$

which is a system of linear equations for $a$ and $b$, and it is straightforward to obtain the solutions as

$$a = \frac{1}{n}[\sum_{i=1}^{n}y_i - b\sum_{i=1}^{n}x_i] = \bar{y} - b\bar{x}, \tag{19.85}$$

$$b = \frac{n \sum_{i=1}^n x_i y_i - (\sum_{i=1}^n x_i)(\sum_{i=1}^n y_i)}{n \sum_{i=1}^n x_i^2 - (\sum_{i=1}^n x_i)^2}, \tag{19.86}$$

where

$$\bar{x} = \frac{1}{n} \sum_{i=1}^n x_i, \quad \bar{y} = \frac{1}{n} \sum_{i=1}^n y_i. \tag{19.87}$$

If we use the following notations

$$K_x = \sum_{i=1}^n x_i, \quad K_y = \sum_{i=1}^n y_i, \tag{19.88}$$

and

$$K_{xx} = \sum_{i=1}^n x_i^2, \quad K_{xy} = \sum_{i=1}^n x_i y_i, \tag{19.89}$$

then the above equation for $a$ and $b$ becomes

$$a = \frac{K_{xx} K_y - K_x K_y}{n K_{xx} - (K_x)^2}, \quad b = \frac{n K_{xy} - K_x K_y}{n K_{xx} - (K_x)^2}. \tag{19.90}$$

The residual error is defined by

$$\epsilon_i = y_i - (a + b x_i), \tag{19.91}$$

whose sample mean is given by

$$\mu_\epsilon = \frac{1}{n} \sum_{i=1}^n \epsilon_i = \frac{1}{n} y_i - a - b \frac{1}{n} \sum_{i=1}^n x_i$$

$$= \bar{y} - a - b\bar{x} = [\bar{y} - b\bar{x}] - a = 0. \tag{19.92}$$

The sample variance $S^2$ is

$$S^2 = \frac{1}{n-2} \sum_{i=1}^n [y_i - (a + b x_i)]^2, \tag{19.93}$$

where the factor $1/(n-2)$ comes from the fact that two constraints are need for the best fit, and the residuals therefore have a $n-2$ degrees of freedom.

## Correlation Coefficient

The correlation coefficient $r_{x,y}$ is a very useful parameter to find any potential relationship between two sets of data $x_i$ and $y_i$ for two random variables $x$ and $y$, respectively. If $x$ has a mean $\mu_x$ and a sample variance $S_x^2$, and $y$ has a mean $\mu_y$ and a sample variance $S_y^2$, the correlation coefficient is defined by

$$r_{x,y} = \frac{\text{cov}(x, y)}{S_x S_y} = \frac{E[xy] - \mu_x \mu_y}{S_x S_y}, \tag{19.94}$$

where $\text{cov}(x, y) = E[(x - \mu_x](y - \mu_y)$ is the covariance. If the two variables are independent $\text{cov}(x, y) = 0$, there is no correlation between them ($r_{x,y} = 0$). If $r_{x,y}^2 = 1$, then there is a linear relationship between these two variables. $r_{x,y} = 1$ is an increasing linear relationship where the increase of one variable will lead to increase of another. $r_{x,y} = -1$ is a decreasing relationship when one increases while the other decreases.

For $n$ sets of data points $(x_i, y_i)$, the correlation coefficient can be calculated by

$$r_{x,y} = \frac{n \sum_{i=1}^n x_i y_i - \sum_{i=1}^n x_i \sum_{i=1}^n y_i}{\sqrt{[n \sum x_i^2 - (\sum_{i=1}^n x_i)^2][n \sum_{i=1}^n y_i^2 - (\sum_{i=1}^n y_i)^2]}}, \tag{19.95}$$

or

$$r_{x,y} = \frac{n K_{xy} - K_x K_y}{\sqrt{(n K_{xx} - K_x^2)(n K_{yy} - K_y^2)}}, \tag{19.96}$$

where $K_{yy} = \sum_{i=1}^n y_i^2$.

---

❑ ***Example 19.7:*** *Is there any relationship between shoe size and height among general population? By collecting data randomly among our friends, we have the following data:*

*Height ($h$): 162, 167, 168, 171, 174, 176, 183, 179 (cm);*
*Shoe size ($s$): 5.5, 6, 7.5, 7.5, 8.5, 10, 11, 12.*

*From these data, we know the sample mean* $\mu_h = 172.5$, $\mu_s = 8.5$.

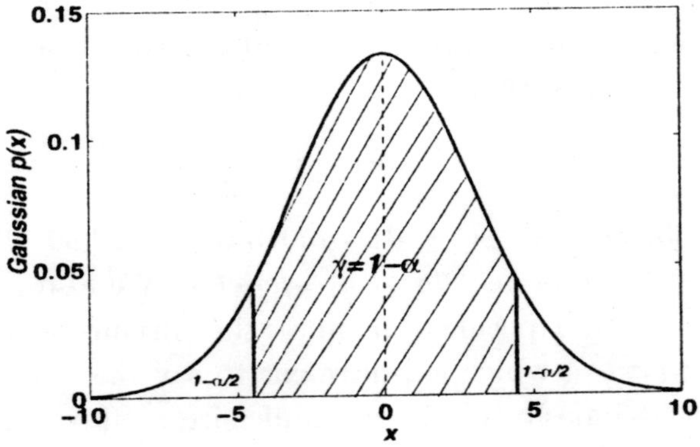


Figure 19.9: Confidence interval $\gamma = 1 - \alpha$.

*The covariance* $\text{cov}(h, s) = E[(h - \mu_h)(s - \mu_s)] = 13.2$. *We also have the standard deviation of height* $S_h = 6.422$ *and the standard deviation of shoe size* $S_s = 2.179$. *Therefore, the correlation coefficient* $r$ *is given by*

$$r = \frac{\text{cov}(h, s)}{S_h S_s} \approx \frac{13.2}{6.422 * 2.179} \approx 0.94.$$

*This is a relatively strong correlation indeed.* □

### 19.2.3 Hypothesis Testing

#### Confidence Interval

The confidence interval is defined as the interval $\theta_1 \leq X \leq \theta_2$ so that the probabilities at these two limits $\theta_1$ and $\theta_2$ are equal to a given probability $\gamma = 1 - \alpha$ (say, 95% or 99%). That is

$$P(\theta_a \leq X \leq \theta_2) = \gamma = 1 - \alpha. \tag{19.97}$$

The predetermined parameter $\gamma$ is always near 1 so that it can be expressed as a small deviation $\alpha \ll 1$ from 1 (see Figure 19.9). If we choose $\gamma = 95\%$, it means that we can expect that about 95% of the sample will fall in the confidence interval while 5% of the data will not.

For the standard normal distribution, this means $P(-\theta \leq \xi \leq \theta) = 1 - \alpha$, so that

$$\phi(\xi \leq \theta) = 1 - \frac{\alpha}{2}. \tag{19.98}$$

If $\alpha = 0.05$, we have $\phi(\xi \leq \theta) = 0.975$ or $\theta = 1.960$. That is to say, $-\theta \leq \xi \leq \theta$ or $\mu - \theta\sigma \leq x \leq \mu + \theta\sigma$. We also know that if you repeat an experiment $n$ times, the variance will decrease from $\sigma^2$ to $\sigma^2/n$, which is equivalent to say that the standard deviation becomes $\sigma/\sqrt{n}$ for a sample size $n$. If $\alpha = 0.01$, then $\theta = 2.579$, we have

$$\mu - 2.579\frac{\sigma}{\sqrt{n}} \leq x \leq \mu + 2.579\frac{\sigma}{\sqrt{n}}. \tag{19.99}$$

On the other hand, for $\theta = 1$, we get $\mu - \sigma \leq x \leq \mu + \sigma$ and $\gamma = 0.682$. In other words, only 68.2% of the sample data will fall in the interval $[\mu - \sigma, \mu + \sigma]$ or

$$x = \mu \pm \sigma, \tag{19.100}$$

with a 68.2% confidence level.

It is conventional to use $\gamma = 0.95$ for probably significant, 0.99 for significant, and 0.999 for highly significant.

---

□ ***Example 19.8:*** *The sample data of the time taken for a quick lunch at a restaurant are as follows: 19, 15, 30, 20, 15, 23, 28, 22, 23 minutes. Suppose you want to attend a lecture at 12:30, at what time you should start your order if you want to take 5% chance of being late? The sample mean is*

$$\mu = \bar{x} = \frac{1}{9}(19 + 15 + 30 + 20 + 15 + 23 + 28 + 22 + 23) = 21.67.$$

*The sample variance is*

$$\sigma^2 = \frac{1}{n-1}\sum_{i=1}^{n}(x_i - \mu)^2 = 26.5,$$

*which gives a standard deviation of $\sigma = 5.15$ minutes. If you are willing to take 5% chance, then $\phi(\xi) = 0.95$, it gives $\xi = 1.645$. So you shall start*

$$x = \mu + \xi\sigma = 30.15,$$

*which is about 30 minutes earlier or at about 12:00.* □

## Student's *t*-distribution

The Student's *t*-test is a very powerful method for testing the null hypothesis to see if the means of two normally distributed samples are equal. This method was designed by W. S. Gosset in 1908 and he had to use a pen name 'Student' because of his employer's policy in publishing research results at that time. This is a powerful method for hypothesis testing using small-size samples. This test can also be used to test if the slope of the regression line is significantly different from 0. It has become one of the most popular methods for hypothesis testing. The theoretical basis of the *t*-test is the Student's *t*-distribution for a sample population with the unknown standard deviation $\sigma$, which of course can be estimated in terms of the sample variance $S^2$ from the sample data.

For $n$ independent measurements/data $x_1, x_2, ..., x_n$ with an estimated sample mean $\bar{x}$ and a sample variance $S^2$ as defined by equation (19.74), the $t$-variable is defined by

$$t = \frac{\bar{x} - \mu}{(S/\sqrt{n})}. \tag{19.101}$$

The Student's $t$-distribution with $k = n - 1$ degrees of freedom is the distribution for the random variable $t$, and the probability density function is

$$p(t) = \frac{\Gamma(\frac{k+1}{2})}{\sqrt{k\pi}\Gamma(k/2)}[1 + \frac{t^2}{k}]^{-\frac{k+1}{2}}. \tag{19.102}$$

It can be verified that the mean is $E[t] = 0$. The variance is $\sigma^2 = k/(k-2)$ for $k > 2$ and infinite for $0 < k \leq 2$.

The corresponding cumulative probability function is

$$F(t) = \frac{\Gamma(\frac{k+1}{2})}{\sqrt{k\pi}\Gamma(k/2)} \int_{\infty}^{t} [1 + \frac{\zeta^2}{k}]^{-\frac{k+1}{2}} d\zeta. \tag{19.103}$$

This integral leads to a hypergeometric function, and it is not straightforward to calculate, that is why they are tabulated in many statistical tables. For a confidence level of $\gamma = 1 - \alpha$, the confidence interval is given by

$$F(\theta) = 1 - \frac{\alpha}{2}, \tag{19.104}$$

which is usually tabulated. For $\alpha = 0.05$ and $0.01$ (or $1 - \alpha/2 = 0.975$ and $0.995$), the values are tabulated in Table 19.2.

Table 19.2: Limits defined by $F(\theta) = 1 - \alpha/2$ in equation (19.104).

| $k$ | $F(\theta)_{0.975}$ | $F(\theta)_{0.995}$ |
|---|---|---|
| 1 | 12.7 | 63.7 |
| 2 | 4.30 | 9.93 |
| 3 | 3.18 | 5.84 |
| 4 | 2.78 | 4.60 |
| 5 | 2.57 | 4.03 |
| 6 | 2.45 | 3.71 |
| 7 | 2.37 | 3.50 |
| 8 | 2.31 | 3.36 |
| 9 | 2.26 | 3.25 |
| 10 | 2.23 | 3.17 |
| 20 | 2.09 | 2.85 |
| 50 | 2.01 | 2.68 |
| 100 | 1.98 | 2.63 |
| $\infty$ | 1.96 | 2.58 |

Suppose we are dealing with the 95% confidence interval, we have $p(-\theta \leq t \leq \theta) = 1 - \alpha = 0.95$ or $p(t \leq \theta) = 1 - \alpha/2 = 0.975$, we have $\theta = t_{\alpha,k} = 12.70(k = 1)$, $4.30(k = 2)$, $3.18(k = 3)$, ..., $2.228(k = 10)$, ..., $1.959$ for $k \to \infty$. Hence,

$$\mu - \theta \frac{S}{\sqrt{n}} \leq t \leq \mu + \theta \frac{S}{\sqrt{n}}. \tag{19.105}$$

This is much more complicated than its counterpart of the standard normal distribution.

### Student's $t$-test

There are quite a few variations of the Student's $t$-test, and most common are the one sample $t$-test and the two sample $t$-test. The one sample $t$-test is used for measurements that are randomly drawn from a population to compare the sample mean with a known number.

In order to do statistical testing, we first have to pose precise questions or form a hypothesis, and such hypothesis is conventionally called the null hypothesis. The basic steps of a $t$-test are as follows:

1. The null hypothesis: $H_0$: $\mu = \mu_0$ (often known value) for one sample, or $H_0$: $\mu_1 = \mu_2$ for two samples;

2. Calculate the $t$-test statistic $t$ and find the critical value $\theta$ for a given confidence level $\gamma = 1 - \alpha$ by using $F(t \leq \theta) = 1 - \alpha/2$;

3. If $|t| > \theta$, reject the hypothesis. Otherwise, accept the hypothesis.

---

□ ***Example 19.9:*** *A group of candidates (say, more than 100 students) have claimed to have an averaged IQ of 110 (or $\mu_0 = 110$). Then, you randomly sample 11 students to do the IQ test and results are: $x = IQ = 106, 112, 103, 108, 108, 109, 100, 106, 106, 99, 101$. Test the hypothesis:*

$$\mathrm{H}_0 : \mu = \mu_0,$$

*at a confidence level of 95%.*

*From the data, we know that $n = 11$, $\bar{x} = 105.273$, $S = 4.077$. Then, we have*

$$t = \frac{(\bar{x} - \mu)}{(S/\sqrt{n})} = \frac{(105.273 - 110)}{4.077/\sqrt{11}} \approx -3.846.$$

We only use the positive value if we look at the statistical tables. We also know for $k = n - 1 = 10$ degrees of freedom at a 95% confidence level; $\theta = 2.228$. At a 95% confidence level, the probability of $t > \theta$ is 0.025 (or 2.5%) and the probability $t < -\theta$ is also 0.025. Thus, the hypothesis is not valid at a 95% confidence level. At the same level of confidence, the true mean $\mu_0$ lies in in the range of $\bar{x} - 2.228 * S/\sqrt{11} \leq \mu_0 \leq \bar{x} + 2.228S/\sqrt{11}$ or

$$102.53 \leq \mu_0 \leq 108.00.$$

□

---

Another important $t$-test is the two-sample paired test. Assuming that two pairs of $n$ sample data sets $U_i$ and $V_i$ are independent and drawn from the same normal distribution, the paired $t$-test is used to determine whether they are significantly different from each other. The $t$-variable is defined by

$$t = \frac{(\bar{U} - \bar{V})}{S_d/\sqrt{n}} = (\bar{U} - \bar{V})\sqrt{\frac{n(n-1)}{\sum_{i=1}^{n}(\tilde{U}_i - \tilde{V}_i)^2}}, \tag{19.106}$$

where $\tilde{U}_i = U_i - \bar{U}$ and $\tilde{V}_i = V_i - \bar{V}$. In addition,

$$S_d^2 = \frac{1}{n-1}\sum_{i=1}^{n}(\tilde{U}_i - \tilde{V}_i)^2. \tag{19.107}$$

This is equivalent to apply the one-sample test to the difference $U_i - V_i$ data sequence.

---

□ ***Example 19.10:*** *A novel teaching method of teaching children science was tried in a class (say class $B$), while a standard method was used in another class (say class $A$). At the end of the assessment, 8 students are randomly drawn from each class, and their science scores are as follows:*
*Class A: $U_i$ = 76, 77, 76, 81, 77, 76, 75, 82;*
*Class B: $V_i$ = 79, 81, 77, 86, 82, 81, 82, 80.*
*At a 95% confidence level, can you say the new method is really better than the standard method?*

*If we suppose that the two methods do not produce any difference in results, that is to say, their mean are the same. Thus the null hypothesis is:*

$$H_0 : \mu_A = \mu_B.$$

*We know that $\bar{U} = 77.5$, $\bar{V} = 81$. The combined sample variance $Sd = 2.828$. We now have*

$$t = \frac{\bar{U} - \bar{V}}{S_d/\sqrt{n}} = \frac{77.5 - 81}{2.828/\sqrt{8}} = -3.5.$$

*We know from the statistical table that the critical value $\theta = 2.37$ for $F(\theta) = 1 - \alpha/2$ and $k = n - 1 = 7$. As $t < -\theta$ or $t > \theta$, we can reject the null hypothesis. That is to say, the new method does produce better results in teaching science.* □

The variance analysis and hypothesis testing are important topics in applied statistics, and there are many excellent books on these topics. Readers can refer to the relative books listed at the end of this book. It is worth pointing out that other important methods for hypothesis testing are Fisher's $F$-test, $\chi^2$-test, and non-parametric tests. What we have discussed in this chapter is just a tip of an iceberg, however, it forms a solid basis for further studies.

# References

Abramowitz M. and Stegun I. A., *Handbook of Mathematical Functions*, Dover Publication, (1965).

Arfken G., *Mathematical Methods for Physicists*, Academic Press, (1985).

Ashby M. F. and Jones D. R., *Engineering Materials*, Pergamon Press, (1980).

Atluri S. N., *Methods of Computer Modeling in Engineering and the Sciences*, Vol. I, Tech Science Press, 2005.

Audoly B. and Neukirch S., Fragmentation of rods by cascading cracks: why spaghetti do not break in half, *Phys. Rev. Lett.*, **95**, 095505 (2005).

Bathe K. J., *Finite Element Procedures in Engineering Analysis*, Prentice-Hall, (1982).

Carrrier G. F. and Pearson C. E., *Partial Differential Equations: Theory and Technique*, 2nd Edition, Academic Press, 1988.

Carslaw, H. S., Jaeger, *Conduction of Heat in Solids*, 2nd Edition, Oxford University Press, (1986).

Chriss N., *Black-Scholes and Beyond: Option Pricing Models*, Irwin Professional Publishing, (1997).

Courant R. and Hilbert, D., *Methods of Mathematical Physics*, 2 volumes, Wiley-Interscience, New York, (1962).

Crank J., *Mathematics of Diffusion*, Clarendon Press. Oxford, (1970).

Devaney R. L., *An Introduction to Chaotic Dynamical Systems*, Redwood, (1989).

Drew, D. A., Mathematical modelling of two-phase flow, *A. Rev. Fluid Mech.*, **15**, 261-291 (1983).

Fenner R. T., *Engineering Elasticity*, Ellis Horwood Ltd, (1986)

Fowler A. C., *Mathematical Models in the Applied Sciences*, Cambridge University Press, (1997).

Farlow S. J., *Partial Differential Equations for Scientists and Engineers*, Dover Publications, (1993).

Fletcher, C. A. J., Fletcher C. A., *Computational Techniques for Fluid Dynamics*, Vol. I, Springer-Verlag, GmbH, (1997)

Gardiner C. W., *Handbook of Stochastic Methods*, Springer, (2004).

Gershenfeld N., *The Nature of Mathematical Modeling*, Cambridge University Press, (1998).

Goldreich P. and S. Tremaine, The dynamics of planetary rings, *Ann. Rev. Astron. Astrophys.*, **20**, 249-83 (1982).

Goodman R., *Teach Yourself Statistics*, London, (1957).

Gleick J., *Chaos: Making a New Science*, Penguin, (1988).

Hinch E.J., *Perturbation Methods*, Cambridge Univ. Press, (1991).

Hull J. C., *Options, Futures and Other Derivatives*, Prentice-Hall, 3rd Edition, (1997).

Jeffrey A., *Advanced Engineering Mathematics*, Academic Press, (2002).

John F., *Partial Differential Equations*, Springer, New York, (1971).

Jouini E., Cvitanic J. and Musiela M., *Handbook in Mathematical Finance*, Cambridge Univ Press, (2001).

Kardestruncer H. and Norrie D. H., *Finite Element Handbook*, McGraw-Hill, (1987).

Keener J., Sneyd J., *A Mathematical Physiology*, Springer-Verlag, New York, (2001).

Korn G. A. and Korn T. M., *Mathematical Handbook for Scientists and Engineers*, Dover Publication, (1961).

Kreyszig E., *Advanced Engineering Mathematics*, 6th Edition, Wiley & Sons, New York, (1988).

Kant T., *Finite Elements in Computational Mechanics*, Vols. I/II, Pergamon Press, Oxford, (1985).

Langtangen, H P, *Computational Partial Differential Equations: Numerical Methods and Diffpack Programming*, Springer, (1999).

LeVeque R. J., *Finite Volume Methods for Hyperbolic Problems*, Cambridge University Press, (2002).

Lewis R. W., Morgan K., Thomas H., Seetharamu S. K., *The Finite Element Method in Heat Transfer Analysis*, Wiley & Sons, (1996).

Mandelbrot B. B., *The Fractal Geometry of Nature*, W.H. Freeman (1982).

Mitchell A. R. and Griffiths D. F., *Finite Difference Method in Partial Differential Equations*, Wiley & Sons, New York, (1980).

Moler C. B., *Numerical Computing with MATLAB*, SIAM, (2004).

Murray J. D., *Mathematical Biology*, Springer-Verlag, New York, (1998).

Ockendon J., Howison S., Lacey A., and Movchan A., *Applied Partial Differential Equations*, Oxford University Press, (2003).

Pallour J. D. and Meadows D. S., *Complex Variables for Scientists and Engineers*, Macmillan Publishing Co., (1990).

Papoulis A., *Probability and statistics*, Englewood Cliffs, (1990).

Pearson C. E., *Handbook of Applied Mathematics*, 2nd Ed, Van Nostrand Reinhold, New York, (1983).

Press W. H., Teukolsky S. A., Vetterling W. T., Flannery B. P., *Numerical Recipe in C++: The Art of Scientific Computing*, 2nd Edition, Cambridge University Press, (2002).

Puckett E. G., Colella, P., *Finite Difference Methods for Computational Fluid Dynamics*, Cambridge University Press, (2005).

Riley K. F., Hobson M. P., and Bence S. J., *Mathematical Methods for Physics and Engineering*, 3rd Edition, Cambridge University Press (2006).

Ross S., *A first Course in Probability*, 5th Edition, Prentice-Hall, (1998).

Selby S. M., *Standard Mathematical Tables*, CRC Press, (1974).

Sicardy B., *Dynamics of Planetary Rings*, Lecture Notes in Physics, **682**, 183-200 (2006).

Strang G. and Fix G. J., *An Analysis of the Finite Element Method*, Prentice-Hall, Englewood Cliffs, NJ, (1973).

Smith, G. D., *Numerical Solutions of Partial Differential Equations: Finite Differerence Methods*, 3rd ed., Clarendon Press, Oxford, (1985).

Thomee V., *Galerkin Finite Element Methods for Parabolic Problems*, Springer-Verlag, Berlin, (1997).

Terzaghi, K., *Theorectical Soil Mechanics*, New York, Wiley, (1943).

Turcotte, D. L. & Schubert, G., *Geodynamics: Application of Continuum Physics to Geological Problems*, 1st. ed., John Wiley, New York, (1982).

Weisstein E. W., http://mathworld.wolfram.com

Wikipedia, http://en.wikipedia.com

Wylie C. R., *Advanced Engineering Mathematics*, Tokyo, (1972).

Versteeg H. K, Malalasekra W., Malalasekra W., *An Introduction to Computational Fluid Dynamics: The Finite Volume Method*, Prentice Hall, (1995).

Yang X. S., Young Y., Cellular automata, PDEs and pattern formation (Chapter 18), in *Handbook of Bioinspired Algorithms*, edited by Olarius S. and Zomaya A., Chapman & Hall / CRC, (2005).

Yang X. S., *An Introduction to Computational Engineering with Matlab*, Cambridge Int. Science Publishing, (2006).

Zienkiewicz O C and Taylor R L, *The Finite Element Method*, vol. I/II, McGraw-Hill, 4th Edition, (1991).

# Appendix A

# Mathematical Formulas

This is a summary for the mathematical formulas that have appeared in various sections in this book. We list these formulas for your easy reference.

## A.1 Differentiations and Integrations

**Differentiation Rules:**

$$(uv)' = u'v + uv'$$

$$(\frac{u}{v})' = \frac{u'v - uv'}{v^2}$$

$$\{f[g(x)]\}' = f'[g(x)] \cdot g'(x)$$

**Leibnitz's Theorem:**

$$\frac{d^n}{dx^n}(uv) = u^{(n)}v + nu^{(n-1)}v' + ... + \begin{pmatrix} n \\ r \end{pmatrix} u^{(n-r)}v^{(r)}$$

$$+ ... + uv^{(n)}, \qquad \begin{pmatrix} n \\ r \end{pmatrix} = \frac{n!}{r!(n-r)!}$$

**Integration by parts**

$$\int_a^b u\frac{dv}{dx}dx = [uv]\Big|_a^b + \int_a^b v\frac{du}{dx}dx$$

**Differentiation of an integral**

$$\frac{d}{dx}\int_{a(x)}^{b(x)} u(x,y)dy = [u(x,b)\frac{db}{dx} - u(x,a)\frac{da}{dx}] + \int_{a(x)}^{b(x)} \frac{\partial u(x,y)}{\partial x}dy$$

**Power Series**

$$e^z = 1 + z + \frac{z^2}{2!} + \ldots + \frac{z^n}{n!}\ldots \quad (z \in \mathcal{C})$$

$$\sin z = z - \frac{z^3}{3!} + \frac{z^5}{5!} - \ldots, \qquad \cos z = 1 - \frac{z^2}{2!} + \frac{z^4}{4!} - \ldots$$

$$\sinh z = z + \frac{z^3}{3!} + \frac{z^5}{5!} + \ldots, \qquad \cosh z = 1 + \frac{z^2}{2!} + \frac{z^4}{4!} + \ldots$$

**Complex Numbers**

$$e^{i\theta} = \cos\theta + i\sin\theta, \qquad [e^{i\pi} + 1 = 0].$$

$$z = x + iy = re^{i\theta} = r(\cos\theta + i\sin\theta)$$

De Moivre's formula:

$$[r(\cos\theta + i\sin\theta)]^n = r^n(\cos n\theta + i\sin n\theta)$$

## A.2 Vectors and Matrices

**Dot Product**

$$\mathbf{a}\cdot\mathbf{b} = |\mathbf{a}||\mathbf{b}|\cos\theta = a_i b_j \delta_{ij} = a_1 b_1 + a_2 b_2 + a_3 b_3$$

**Cross Product**

$$\mathbf{a}\times\mathbf{b} = \mathbf{n}|\mathbf{a}||\mathbf{b}|\sin\theta = \epsilon_{ijk}a_j b_k = \begin{vmatrix} \mathbf{i} & \mathbf{j} & \mathbf{k} \\ a_1 & a_2 & a_3 \\ b_1 & b_2 & b_3 \end{vmatrix}$$

**Vector Triple**

$$[\mathbf{a}, \mathbf{b}, \mathbf{c}] \equiv \mathbf{a} \cdot (\mathbf{b} \times \mathbf{c}) = \begin{vmatrix} a_1 & a_2 & a_3 \\ b_1 & b_2 & b_3 \\ c_1 & c_2 & c_3 \end{vmatrix}$$

$$\mathbf{a} \cdot (\mathbf{b} \times \mathbf{c}) = \mathbf{b} \cdot (\mathbf{c} \times \mathbf{a}) = \mathbf{c} \cdot (\mathbf{a} \times \mathbf{b}) = -\mathbf{a} \cdot (\mathbf{c} \times \mathbf{b})$$

$$\mathbf{a} \times (\mathbf{b} \times \mathbf{c}) = (\mathbf{a} \cdot \mathbf{c})\mathbf{b} - (\mathbf{a} \cdot \mathbf{b})\mathbf{c}$$

**Divergence Theorem of Gauss**

$$\iiint_V \nabla \cdot \boldsymbol{u} dV = \iint_S \boldsymbol{u} \cdot d\mathbf{S}$$

**Stokes's Theorem**

$$\iint_S (\nabla \times \boldsymbol{u}) \cdot d\mathbf{S} = \oint_\Gamma \boldsymbol{u} \cdot d\boldsymbol{\Gamma}$$

**Green's Theorems**

$$\int_V (\psi \nabla^2 \phi - \phi \nabla^2 \psi) dV = \int_S (\psi \frac{\partial \phi}{\partial n} - \phi \frac{\partial \psi}{\partial n}) dS$$

$$\oint (u dx + v dy) = \iint (\frac{\partial v}{\partial x} - \frac{\partial u}{\partial y}) dx dy$$

**Identities**

$$\nabla \cdot \nabla \times \boldsymbol{u} = 0, \quad \nabla \times \nabla \phi = 0$$

$$\nabla \times (\phi \mathbf{u}) = \phi \nabla \times \mathbf{u} + (\nabla \phi) \times \mathbf{u}$$

$$\nabla \cdot (\phi \mathbf{u}) = \phi \nabla \cdot \mathbf{u} + (\nabla \phi) \cdot \mathbf{u}$$

$$\nabla \times (\nabla \times \mathbf{u}) = \nabla (\nabla \cdot \mathbf{u}) - \nabla^2 \mathbf{u}$$

**Inverse, Trace and Determinants**

$$(\mathbf{AB...Z})^{\mathbf{T}} = \mathbf{Z}^{\mathbf{T}}...\mathbf{B}^{\mathbf{T}}\mathbf{A}^{\mathbf{T}}, \quad (\mathbf{AB...Z})^{-1} = \mathbf{Z}^{-1}...\mathbf{B}^{-1}\mathbf{A}^{-1}$$

$$|\mathbf{AB....Z}| = |\mathbf{A}||\mathbf{B}|...|\mathbf{Z}|, \quad |\mathbf{A}| = \det \mathbf{A}$$

$$\mathbf{A}\mathbf{v}_i = \lambda_i \mathbf{v}_i, \quad \mathrm{eig}(\mathbf{AB}) = \mathrm{eig}(\mathbf{BA})$$

$$\mathrm{tr}(\mathbf{A}) = \sum_i \mathbf{A}_{ii} = \sum_i \lambda_i, \qquad \det(\mathbf{A}) = \Pi_i \lambda_i$$

$$\mathrm{tr}(\mathbf{AB}) = \mathrm{tr}(\mathbf{BA}), \qquad \mathrm{tr}(\mathbf{A}+\mathbf{B}) = \mathrm{tr}(\mathbf{A}) + \mathrm{tr}(\mathbf{B})$$

$$\det(\mathbf{A}^{-1}) = \frac{1}{\det(\mathbf{A})}, \qquad \det(\mathbf{AB}) = \det(\mathbf{A})\det(\mathbf{B})$$

**Exponential Matrices**

$$e^{\mathbf{A}} \equiv \sum_{n=0}^{\infty} \frac{1}{n!}\mathbf{A}^n = \mathbf{I} + \mathbf{A} + \frac{1}{2}\mathbf{A}^2 + ...$$

$$e^{t\mathbf{A}} \equiv \sum_{n=0}^{\infty} \frac{1}{n!}(t\mathbf{A})^n = \mathbf{I} + t\mathbf{A} + \frac{t^2}{2}\mathbf{A}^2 + ...$$

$$\ln(\mathbf{IA}) \equiv \sum_{n=1}^{\infty} \frac{(-1)^{n-1}}{n!}\mathbf{A}^n = \mathbf{A} - \frac{1}{2}\mathbf{A}^2 + \frac{1}{3}\mathbf{A}^3 + ...$$

$$e^{\mathbf{A}}e^{\mathbf{B}} = e^{\mathbf{A}+\mathbf{B}} \qquad (\text{if } \mathbf{AB} = \mathbf{BA})$$

$$\frac{d}{dt}e^{t\mathbf{A}} = \mathbf{A}e^{t\mathbf{A}} = e^{t\mathbf{A}}\mathbf{A}$$

$$(e^{\mathbf{A}})^{-1} = e^{-\mathbf{A}}, \qquad \det(e^{\mathbf{A}}) = e^{\mathrm{tr}\mathbf{A}}$$

## A.3 Asymptotics

**Gaussian Distribution**

$$p(x;\mu,\sigma) = \frac{1}{\sqrt{2\pi\sigma^2}}\exp[-\frac{(x-\mu)^2}{2\sigma^2}] \to \delta(x) \qquad \sigma \to \infty.$$

$$\int_{-\infty}^{\infty} e^{-\alpha x^2} dx = \sqrt{\frac{\pi}{\alpha}}$$

$$\int_{-\infty}^{\infty} x^{2n} e^{-\alpha x^2} dx = \frac{(-1)^n \cdot 1 \cdot 3 \cdots (2n-1)}{2^n}\sqrt{\frac{\pi}{\alpha^{2n+1}}}, \ (n > 0).$$

**Binomial Distribution**

$$B(k;n,p) = \frac{n!}{(n-k)!k!}\, p^k(1-p)^{n-k}, \ (k = 0, 1, 2, ..., n)$$

$$B(x; n \to \infty, p)\Big|_{np\gg 1} \sim p(x; \mu, \sigma), \ \mu = np, \ \sigma^2 = np(1-p).$$

**Poisson Distribution**

$$f(x; \lambda) = \frac{e^{-\lambda}\lambda^x}{x!}, \qquad x \in \mathcal{N}$$

$$f(x; \lambda \gg 1) \approx p(x; \mu, \sigma), \qquad \mu = \lambda, \ \sigma^2 = \lambda$$

$$B(k \to x; n \to \infty, p) \sim f(x; \lambda = np), \qquad (\lim_{n\to\infty} np = \lambda)$$

## A.4 Special Integrals

**Gamma Function**

$$\Gamma(x) = \int_0^\infty t^{x-1}e^{-t}dt = \int_0^\infty e^{-t+(x-1)\ln t}dt$$

$$\Gamma(n+1) = n!, \ \Gamma(-\frac{1}{2}) = -2\sqrt{\pi}, \ \Gamma(\frac{1}{2}) = \sqrt{\pi}$$

$$\Gamma(x+1) \approx (\frac{x}{e})^x\sqrt{2\pi x}, \qquad (x \to \infty).$$

**Stirling's Formula**

$$n! \approx (\frac{n}{e})^n\sqrt{2\pi n}, \qquad n \gg 1$$

**Error Functions**

$$\operatorname{erf}(x) = \frac{2}{\sqrt{\pi}}\int_0^x e^{-\eta^2}d\eta \sim 1 - \frac{e^{-x^2}}{x\sqrt{\pi}}, \ (x \to \infty)$$

$$\operatorname{erfc}(x) = 1 - \operatorname{erf}(x) = \frac{2}{\sqrt{\pi}}\int_x^\infty e^{-t^2}dt$$

$$\operatorname{erf}(x) \sim \frac{2}{\sqrt{\pi}}[x - \frac{x^3}{3} + \frac{x^5}{10} - \ldots], \ (x < \infty).$$

# Index